THE DO·IT·YOURSELF WEATHER BOOK

To Carolyn and Kristi

CONTENTS

ILLUSTRATIONS

Figure

Figure

Figure

ACKNOWLEDGEMENTS

No book is ever an individual effort. Some contributions come indirectly from chance acquaintances and experiences which spawn ideas or redirect the course of your thoughts. Other contributions come directly from sustained contact with individuals who possess essential specialized knowledge and who are willing to save you literally years of research by sharing this knowledge. Finally, there are the efforts by a few key people who help to construct, edit, and polish the end results of your toil to produce a vehicle which will accurately reflect what you are trying to communicate.

The first individuals to be acknowledged should be my wife Carolyn and daughter Kristi. Their patience with me as I labored on this book for seemingly endless hours, days, and months bordered on the saintly. They also willingly persevered over the last several years through the trials, tribulations, and uncertainties of starting a new business. In addition, Carolyn personally typed every word of the first draft.

The philosophy expressed in this book is my own but the major portion of it has been shaped by my esteemed associate Lowell Semans. After working with and studying under Lowell for over two years, I have the utmost respect for his ability and character. I, personally, consider him to be the finest practical research meteorologist in the nation and one of the best long-range forecasters in the world. Someday every meteorologist in the world will use the superior forecasting techniques Lowell has developed in nearly four decades of practical experience.

Many other people contributed to this book, not the least of whom was my editor, Mary Jean Haddin. Her tireless devotion to the efficient completion and strategic enhancement of my original manuscript has made the completed product something of which to be proud.

A very special thanks goes to Bill Mainord, E.F. Hutton & Co., who patiently guided me through the intricacies of crop marketing. His advice concerning the technical aspects of crop marketing were invaluable to the completion of chapter six.

Other contributors, in alphabetical order include: D.V. Craven, Tom Curl, Dee Gandy, Jim Harding, Fred Kelly, Nick Korn, Dr. James McQuigg, Tommy Trantham, John Weems, Billy Wooten, and all my farming customers.

INTRODUCTION

You have often heard the statement that everyone talks about the weather but no one ever does anything about it. That statement has, in fact, been outdated for some time now. This book will hopefully be the crest of a tidal wave of efforts to prove just the opposite and to substantiate, once and for all, that a great deal is being done about the weather and that a great deal more will be done in the future.

The purpose of this book is to show that something is being done about the effect of weather on farming and that there is a great deal a farmer can do to actually manage the effect of weather on his land and crops. This book contains a description of a working system for managing the effect of weather on farms, a system which is currently producing results similar to those described in chapters five and six.

The concept is unique for, to my knowledge, it is the only one of its sort in the world. It is also revolutionary for it completely denies all current thoughts about meteorology held by the majority of people in this country—i.e., that absolutely nothing can be done about the weather, so why bother. That is false. In fact, not only can something be done now but it could just as easily have been accomplished long before now.

One of the reasons for writing Part One of this book was the hope of awakening the interest of general readers to some of the fascinating aspects of weather and climate that might be unknown to them. I have attempted to dispel the myth that we meteorologists are incompetent bunglers who are wrong more often than we're right and that we can't communicate with the general public. That myth has been promoted by the modern age of television and radio and its flippant attitude towards weather information of true quality, substance, and importance to society, in favor of the inane weather programs produced by the majority of radio and television stations. The competence of professional meteorologists is slandered each time another beautiful person and/or deep-voiced announcer, instead of a professional meteorologist, gives the latest weather forecast or information on radio or television.

By the time you complete Part One, you'll have an idea of what a professional meteorologist or weather forecaster is, what he does, how he does it, and how well he does it. You will know what to expect and, most important of all, where to find it. Or, if you can't get quality weather information in your area, you'll know who to contact and what you should ask for in order to get what you want.

The only "bad" thing about Part One is that it will raise your expectations by revealing what you should be getting but probably aren't. For example, most people just naturally assume that a TV weatherman knows something about weather. However, most don't, even though they go to great lengths to prove otherwise. I have seen a TV weather announcer who knew nothing about weather forecasting brand himself as the "most renowned weather forecaster" in his area! You will learn how to spot these people and how to look for the professional meteorologist who actually knows what he is talking about. Every community in America, not just the agricultural ones, should have at least one experienced weather forecaster on both radio and television to insure that there is one credible source serving the many and various weather needs of that community.

Every time an unexpected blizzard strands thousands of motorists, stalls hundreds of trucks, brings cities to a standstill, and causes millions of dollars worth of damage, you should ask yourself why these people didn't know about this storm in advance. In most cases, an experienced weather forecaster could have prevented much of the above situation by giving accurate advance notice of the expected storm several days before it actually occurred. These "old pros," mostly retired military or National Weather Service forecasters with superb experience, are available by the score but nobody seems very interested in them. Their collective forecasting ability is potentially worth billions of dollars to industry, agriculture, and the general public. Perhaps, after Part One becomes common knowledge, enough people will raise a big enough ruckus to see that this vast reservoir of brain power and ability is properly harnessed to the good of all.

You will learn the facts about long-range weather forecasting in the first part of the book. You will see that long-range forecasts can be quite accurate and that they are good enough to provide you with usable answers. They will get even better in the future but you've been fooled into thinking that they're not good enough, yet, to use for any intelligent purpose.

Part One is for everyone. Whether you're just an interested bystander or involved with some weather-dependent business, weather affects you directly and will cost you an untold amount of time and money in your lifetime. You need the information in Part One to help you to make better decisions with regard to weather. In some businesses, like trucking, advance knowledge of weather can save hundreds of thousands of dollars in a single day. Most individuals or companies have no idea where to look for weather information or what they can reasonably expect when they do find it. The credibility gap between the public and the professional weather forecaster or meteorologist is enormous.

Part Two is primarily directed to the American farmer. This section contains specific directions about how to apply weather management to farming. It explains in detail how to monitor, analyze, and correct the effects of weather on any farm.

Despite the emphasis on forecasting in the first part of the book, the second part stresses the fact that analyzing where you've been and where you are is just as important as attempting to determine where you're going. To that end, the American farmer must begin keeping accurate records on the effect of weather on his farm. Without these records it is extremely difficult to analyze the effect of weather on his profit margin or to make effective use of any weather forecast. The data he records will provide a solid basis for analysis as well as projection of both weather and crop yields.

The important point to make is that the effects of weather on a given farm are not random. They can be specifically quantified and used to your advantage. Very few farmers realize this fact and almost none know how to take advantage of it. Chapter five will describe how you can quantify the effect and save a small fortune each year by taking advantage of your information.

In chapter six, the complex process of crop marketing is addressed to show that weather plays a vital role in the process and can similarly be used to your advantage. Marketing crops is difficult enough without ignoring the single most important factor effecting both supply and demand—i.e., weather. This effect can be monitored, analyzed, and used to your direct advantage. It's easy to learn how and money in the bank if you'll just take the time.

The conclusions are an important ingredient of this book because I've used them as a sounding board for all that's wrong in the communication, knowledge, and use of weather information in America. However, I don't leave it at that. I state that something can and is being done about the situation and what you can personally do about it. I describe a solution.

There will no doubt be those critics, perhaps even fellow professionals, who will take issue with my solution. They will question its scope, direction, and intent. But let them at least acknowledge that a solution is being attempted to the oldest and most prevalent problem in agriculture—i.e., weather. Let them further note that this problem, which is costing agriculture literally billions of dollars annually, has been addressed only piecemeal in this country, despite the fact that virtually every facet of my concept of weather management, as it applies today, could just as easily have been employed more than half a century ago. Let all who read this then, accept the fact that a viable, working solution has been developed, is in use today, and is working as described herein. It may not be the best possible solution but it is a solution. Enhancements will come but the value of weather management to any farmer who employs it now will be more than sufficient immediately and the dividends through continued use and further enhancements will be just that much greater.

PART I

LEARNING ABOUT WEATHER

1
THE SCIENCE OF WEATHER FORECASTING

An anonymous wit once stated that "forecasting is like driving a car blindfold while following directions from someone looking out the rear window." Perhaps to the casual observer, weather forecasting lives up to that reputation.

In frequent talks to various organizations I discuss the science of weather, and during these talks I always swear my audience to secrecy before revealing such forecasting techniques as the accuracy method (I show a picture of a dartboard), the statistical method (here I show a man flipping a coin), and the Great Equalizer (a rabbit's foot).

Weather forecasting may seem somewhat mysterious to the uninitiated, for the meteorologist is often jokingly accused of actually using dartboards, coins, and rabbits' feet in preparing forecasts. I admit that in the eyes of the general public we meteorologists have a somewhat tarnished image, but it is largely undeserved. Just like any business or profession, once you understand the special set of inherent problems, you will easily see that our collective forecasting ability is much better than you suspect.

Weather forecasting is an ancient pastime, for it has been practiced, with varying degrees of sophistication, since man first looked to the sky. The natural weather-forecasting instincts of animals and insects date back even farther. So it is little wonder that homemade forecast techniques abound and that many of these are related to habits of animals and insects.

Other methods are based on thousands of years of practical experience in watching the weather and looking for changes in cloud cover, wind, temperature, humidity, and other variables that usually signal a change in the weather. There are also changes in plant growth and development that can signal a change in the weather (such as thick corn husks), or record ancient weather patterns (as shown in the thickness of growth rings in very old trees).

There is nothing wrong with using signs in nature to indicate expected weather but it takes time to learn all of the many variations involved. The potential changes are usually quite local and the actual indicators can fool you sometimes. There are very few local records kept of natural phenomena, especially for the animal, plant, and insect worlds, which are consistently good predictors of weather. A partial list of some commonly observed natural phenomena (and the changes in the weather they are presumed to indicate) is given in appendix E.

Of course one of the best natural indicators of expected changes in the weather is weather itself. All of us observe changes in the weather and if you study them long enough you'll begin to see a pattern of changes. For example, if you'll study changes in the clouds, you'll soon begin to observe that high or cirrus clouds are usually followed by middle or altocumulus clouds, then low or cumulus clouds, then rain. Other indications of change are given by changes in weather elements such as temperature, wind, humidity, etc. You can monitor these changes with simple weather instruments you can buy or make yourself (see appendix B). Here again, though, these changes are quite localized and they too can fool you sometimes.

The professional forecaster must be aware of local weather patterns—it never hurts to look out the window! However, weather forecasters will confirm that many a forecast has been ruined simply because the forecaster looked out the window at clear skies in the morning, forecasted a sunny day, and walked home in the rain without an umbrella! So it should come as no surprise to know that, more often than not, we'll rely almost totally on our charts, teletype reports, radar, and statistical data without even bothering to look out the window for fear that it will bias our forecast. We generally practice a self-enforced ignorance of the environment because of modern requirements which necessitate the preparation of forecasts for increasingly larger or more distant areas and not because we have any particular disdain for heeding the signs of nature.

THE BEGINNING OF MODERN FORECASTING

Modern forecasting techniques began in the 1800s with a relatively efficient system of telegraphed observations to a central location where a forecaster would analyze the data. The prepared forecast would then be distributed by mail, newspaper, and back out through the telegraph network. In the late 1800s telegraphed weather observations were taken at regular intervals in major cities, primarily east of the Mississippi River, and relayed to Signal Service headquarters (the old Weather Bureau was born in 1891 as a civilian agency under the U.S. Department of Agriculture) in Washington, D.C., where one man would call out the reports to clerks nearby who would record certain types of weather data. Metal plates with prepared type and individual slots for each observing site would then be updated with the correct symbols. While this was happening, the forecaster would use the data to prepare his forecast. Within two hours the presses were rolling and the new map was printed and distributed by mail.

These original forecasts were usually good for only the next day but could be extended to two to three days on rare occasions. Forecasting, back then, consisted of observing how weather systems were affecting one part of the country, and projecting this effect on subsequent locations based on the assumption that storms generally move from west to east. Little was known of upper-air data and the jet stream was yet to be discovered some 50 years hence.

In 1871 Congress appropriated $15,000 to run the National Weather Service which at that time belonged to the Signal Service.

Today, in place of one man and a set of clerks, the National Meteorological Center in Suitland, Maryland, uses a massive computer to process over 40,000 daily surface observations, plus thousands of ship and aircraft reports and upper-air data. The National Weather Service budget for 1978 was $195 million—over 10,000 times larger than in 1871.

The modern forecaster receives his weather information from the National Meteorological Center via teletypes, radar images, satellite photos, surface and upper-air maps, and computer projection charts. Not only has the quantity of information increased phenomenally but the techniques of forecasting have likewise undergone drastic revision and modernization as a result of the explosion in communication, information, and computer analysis. The emphasis on forecasting has also shifted from complete dependence on surface reports to the increasingly important upper-atmospheric analysis and projection.

TOOLS OF THE TRADE

Before discovering the innermost secrets of weather forecasting, let's first take a look at the tools of the trade. It all starts with the observation site, which consists of the instruments used to measure and record the actual weather data, such as a barometer (pressure), thermometer (temperature), anemometer (wind speed), wind vane (wind direction), and rain gauge.

In the 1930s, scientists developed radiosondes—small instrument packages designed to be sent aloft in the upper atmosphere to measure temperature, humidity, pressure, and wind speed and direction. Twice a day radiosonde balloons are released from hundreds of stations in the U.S. and Canada and tracked by radar to altitudes as high as 10 to 15 miles. This information has completely revolutionized forecast methods.

The latest innovation in weather collection techniques has been the advent of weather satellites in the 1960s. Satellite pictures are a valuable source of weather information and can produce some rather spectacular pictures. The availability of satellite data is still limited and quite expensive for most forecasters but it will eventually become an indispensable tool to any forecast operation.

These are the basic tools, all of which are used to construct the periodic weather reports and maps which come into the forecaster's office 24 hours a day. The maps and periodic weather reports are to the forecaster what body temperatures, chemical analyses, and X-rays are to a doctor.

But we're not quite ready to prepare a forecast until we organize this mass of data. Most of this is done at the National Meteorological Center. Hourly weather observations arrive in the forecaster's office via teletype and help us to pinpoint the actual weather conditions. A typical hourly weather observation looks like this:

KMEM 50BKN 100OVC 6 R 119/78/72/0608

The report basically says that Memphis has cloudy skies with rain, a temperature of 78°, and a northeasterly wind.

The final step is the chart display. A few of the charts include the surface map, (similar to the TV weather map) radar summary (a summary of radar observations from across the United States), weather depiction (a summary of clouds, precipitation, and visibility from across the United States), a satellite photograph, various upper-air charts (these show flow of wind at various levels above the earth's surface), computer analysis charts (these show the expected future flow of wind at various levels above the earth's surface), and more.

We're almost ready to prepare a forecast, but first let me explain that not all forecasters start in the same place, apply the same standards, use the same techniques, or subscribe to the same theories; consequently, we rarely reach exactly the same conclusions on any given day although the differences are usually slight. However, the longer the forecast the greater the disagreement, for the ability to make consistent long-range forecasts—periods more than 5 days in advance—is possessed by few forecasters in America. (More about this in chapter four.)

BACKGROUND FOR A FORECAST

Now, let's prepare a forecast. The first step is to arrive at the office before dawn (if you'd care to quit now I wouldn't blame you!) The next step is to tear and file teletype reports and hourly observations and arrange your chart display. Now we're ready to get down to business by taking a good, long look at upper-air charts from several different altitudes in the atmosphere. We're looking for upper-level ridges or high pressure centers—usually good weather—and upper-level troughs or low pressure centers—usually bad or inclement weather (see figure 1).

We're looking for the same kind of high and low pressure systems that you see on the TV weather map. Once we've found these features we must analyze their most likely movement or development and this is based on one of the most important, but least understood, phenomena in meteorology: the jet stream.

Jet stream winds were not discovered until about 1938 during a high-altitude flight over the Mediterranean by a German Luftwaffe aircraft. By definition, jet stream winds are a band of high-altitude winds which circle the globe. There is more than one jet stream but the dominant jet is usually the Mid-latitude Jet which flows from west to east across the continental U.S. at speeds varying from as low as 30 to 40 m.p.h. in the summer to over 100 m.p.h. in the winter. The average position of the Mid-latitude Jet also varies from southern Canada in the summer to as far south as California, Texas, or Florida in the winter.

The jet stream winds usually determine both the track and intensity of a storm. The trick, then, is to forecast what pattern the jet stream winds will take at the point in the future for which you are forecasting. It is possible to make highly accurate forecasts for a two- to three-day period as much as two weeks in advance. It is not an easy process and requires a great deal of experience but when you learn how to forecast future jet stream patterns correctly, you can make some rather astounding forecasts.

Now it's time to get back to those upper-air charts we were looking at. Once we've determined the future shape of the jet stream we'll be able to tell how strongly and in which direction the low's and high's on our upper-level charts will develop. The jet stream is the general dividing line between cold and warm air and the storm track underneath which weather systems will move. The charts for lower levels in the atmosphere help us to fine-tune the forecast since it is usually at these levels that more local producers of weather will develop. For example, in figure 2, the position of the jet stream will provide you with an idea of the general pattern of

FIGURE 1. WIND FLOW ABOVE THE EARTH'S SURFACE

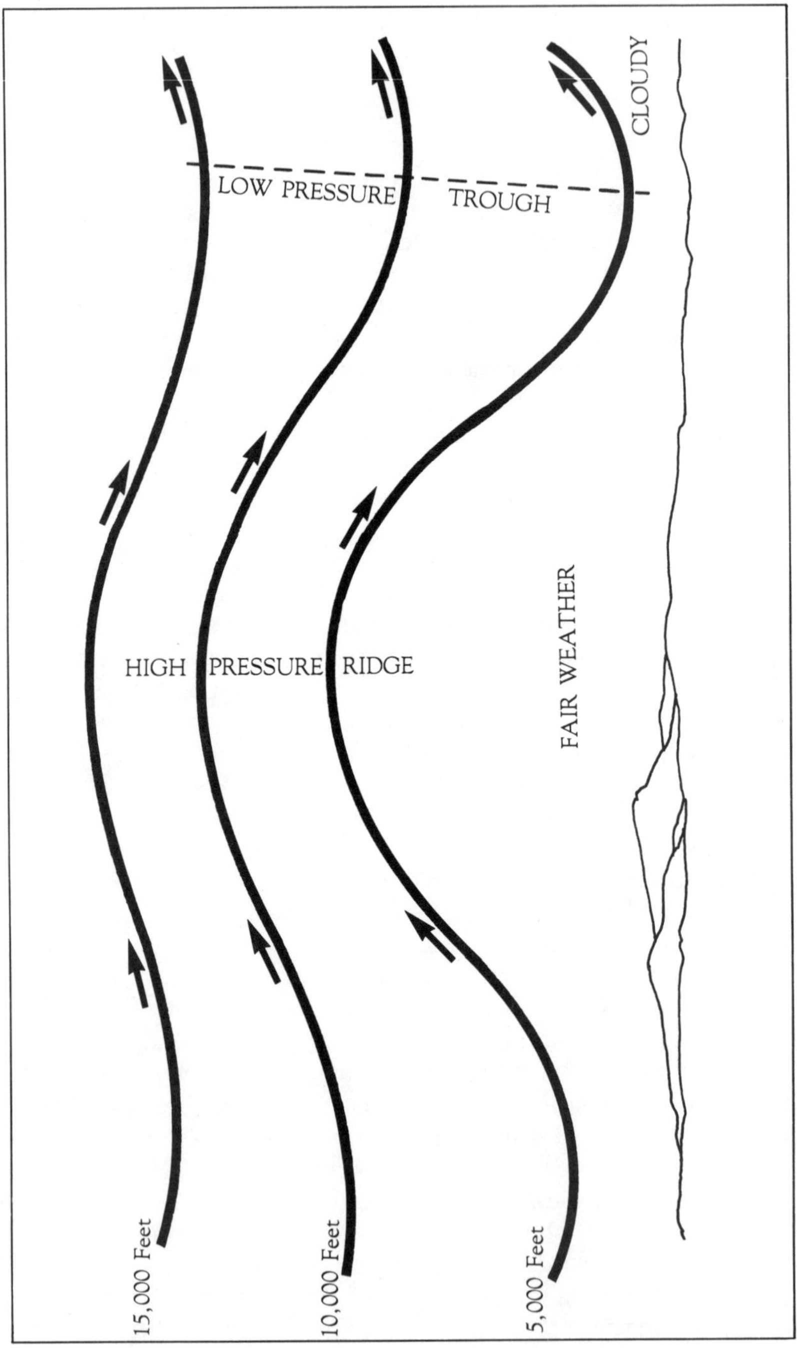

15,000 Feet

10,000 Feet

5,000 Feet

HIGH PRESSURE RIDGE

LOW PRESSURE TROUGH

FAIR WEATHER

CLOUDY

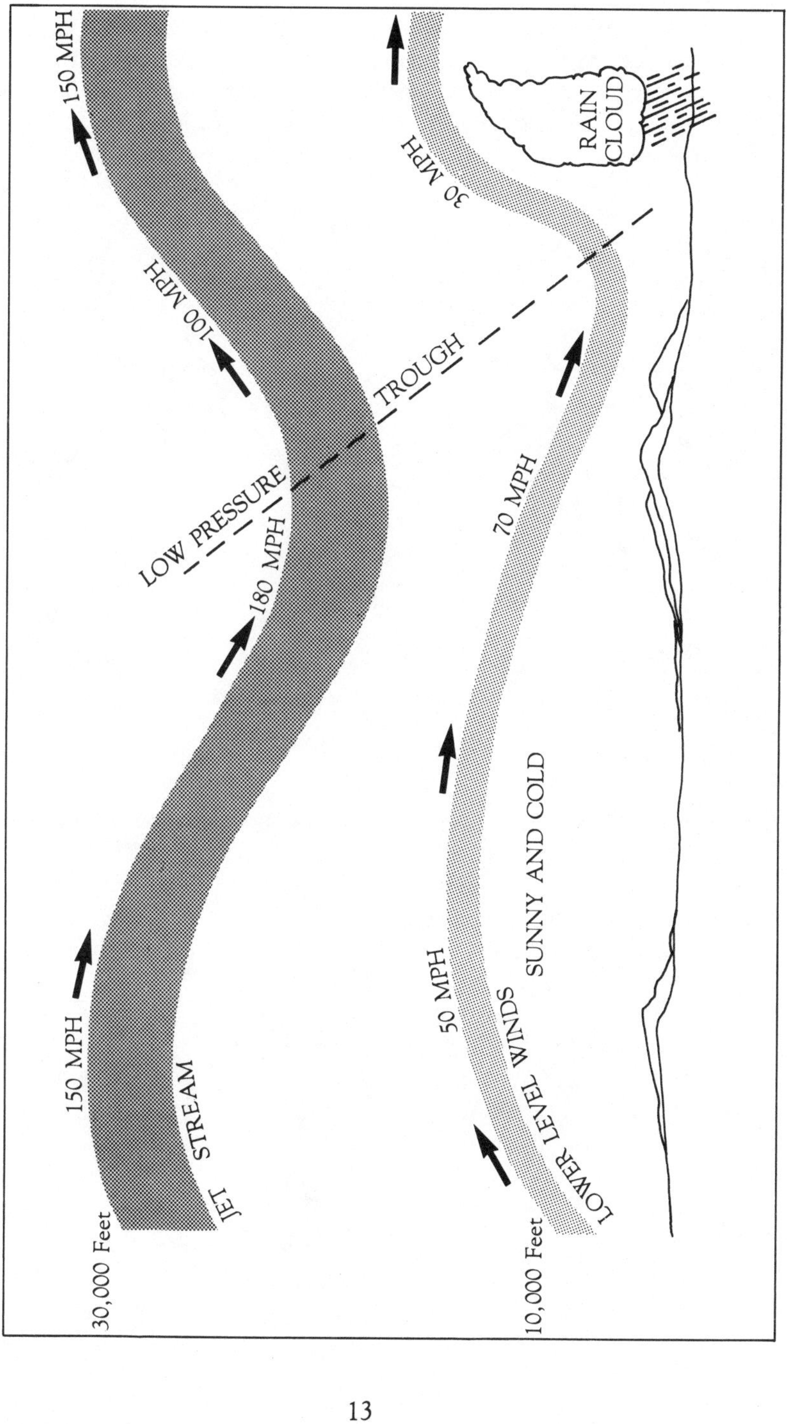

FIGURE 2. RELATIONSHIP BETWEEN JET STREAM AND LOWER LEVEL FLOW

weather to be expected, i.e., colder/warmer or drier/wetter than normal. Once this determination has been made, the 10,000 foot winds will be used to determine the specific weather you can expect for your location. Since rain clouds normally form at the 10,000 foot level, we can use the amount of moisture at that level along with the actual pattern of winds to

FIGURE 3

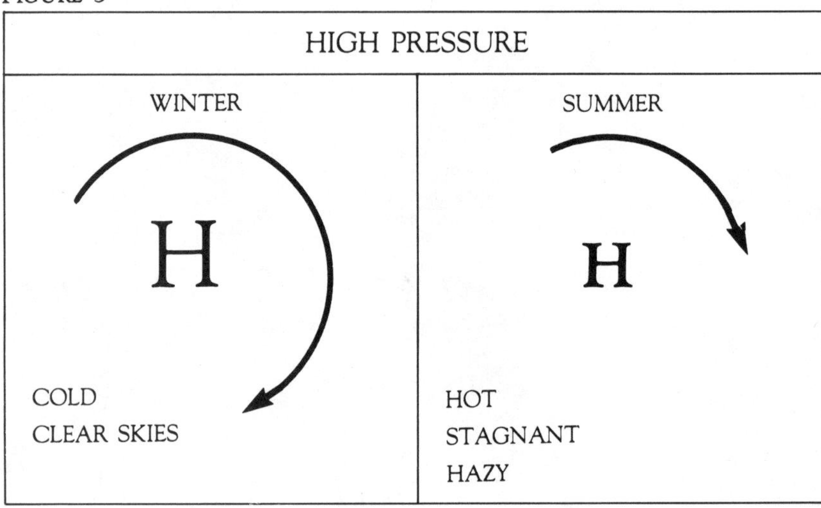

Figure 3.
The circulation around a high pressure cell is similar in both winter and summer, but less intense during the summer.

FIGURE 4

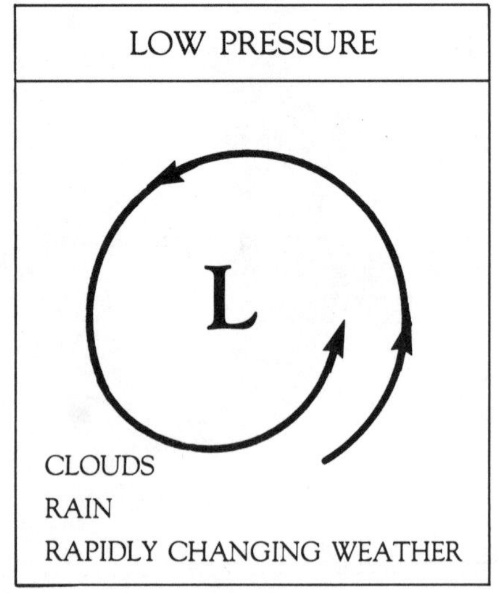

Figure 4.
The circulation around low pressure centers usually results in at least cloudy skies and, normally, rain. Precipitation associated with low pressure centers in the winter covers a given area almost uniformly.

get an excellent idea of the coverage and intensity of precipitation expected. The temperature determination, as made from the jet stream pattern, will help you to determine what type of precipitation to expect, i.e., rain, snow, sleet, freezing rain, etc.

You will notice that I have not yet mentioned the surface weather map. This was intentional for, in most cases, the upper-level wind flow will determine the major direction of the day's forecast. Knowledge of this is then combined with the dominant weather patterns at the earth's surface and all possible local factors before the final forecast is issued.

Now it's time to focus our attention on the surface weather patterns and attempt to discern what large-scale features we'll be dealing with for today and the next several days. A few of these ingredients have already been mentioned, but let's take a little closer look at some of the basic weather producers at the earth's surface.

Let's start with the weather symbols you see on the TV weather map. The most basic symbols are highs and lows. As a general rule of thumb, high pressure systems in fall, winter, and spring are stronger than summer highs and they usually mean colder air and clearer skies with strong winds in advance of the high and calm winds at the center. The clockwise (anticyclonic) flow around a summer high is usually much weaker and the air associated with it much warmer, more stagnant, and more likely to produce hazy skies. A good example of the effects of a summer high is when the Bermuda High develops westward from the Atlantic up over the Southeast resulting in hot, humid, and dry weather for weeks on end.

If high pressure usually connotes fair weather (see figure 3), it follows that low pressure spawns foul weather (see figure 4). Here again, lows are generally stronger in the fall, winter, and spring than in the summer since the counter-clockwise (cyclonic) flow of air around a low is usually stronger in winter than in the summer. Lows can be a grievous source of aggravation to the forecaster when they become detached from the general flow of the jet stream and just drift around in no particular direction, for several days at a time, producing rain and overcast skies.

Low pressure centers are most frequently found attached to the top of a frontal system with its various components: the cold front, the warm front, and the occluded front. The effect of the various frontal components on the weather is quite complex, although some general rules can be applied. For example, the air behind a cold front is usually colder and drier than the air out ahead of it (see figure 5). As the cold air shoots into the warmer and more moist air, the lighter warmer air rises and the moisture condenses into clouds. The faster the cold air moves and the deeper it is, the more likely it is to produce significant shower activity, usually with thunderstorms and, quite frequently, severe weather.

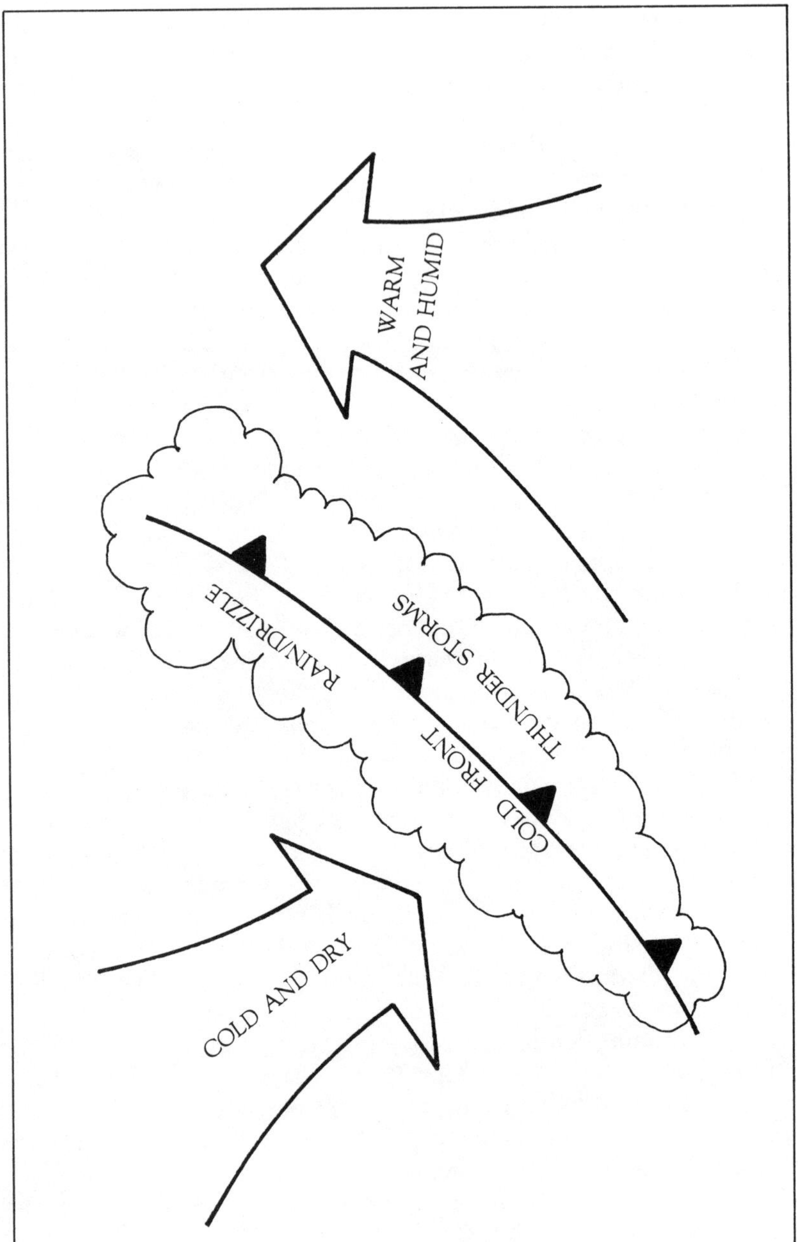

Figure 5. Cold Front

The air behind a warm front (see figure 6) is usually warmer and more moist than the air in front and hence, being lighter, will rise over the cooler air to form clouds. Warm frontal weather is usually less violent than that associated with a cold front and will normally result in less sharply defined clouds and more general rain.

The occluded front is a composite of the two fronts, formed as a cold front overtakes a warm or stationary front (see figure 7). Having components of both warm and cold fronts, the occluded front will usually possess weather common to both, or that found around the low pressure center. The above three components and the low pressure center are usually seen attached to each other in the manner shown in figure 8.

The frontal system in figure 8 is the most common frontal arrangement seen on a weather map. The combination of frontal systems (cold, warm, and occluded) will produce an entire range of weather not only common to each individual front but also possessing a characteristic pattern peculiar to this combination. For example, a frontal system of the type in figure 8 will usually produce a pattern of rain and thunderstorms in front of or coincident with the cold front, a large area of overcast skies with stratus clouds and warm rain in the vicinity of the warm front, and a large area of low stratus and cumulus clouds with general rain for a several hundred-mile radius of the low pressure center.

A high pressure center is included in figure 8 to indicate its normal relation to the typical frontal system as shown. Not all frontal systems are this classic in construction; but especially during the fall, winter, and spring, the majority of frontal systems moving across North America will be at least roughly similar to that shown. It is interesting to note that frontal depiction on weather maps is a relatively recent occurrence, for fronts were not commonly drawn on a weather map until around 1936. The actual application of fronts in forecasting is tied to the analysis of the various air masses that penetrate North America and the usual characteristics of these air masses. (More about this in chapter four.)

Before proceeding to the final step in preparing a forecast, let's take a brief look at the association between the upper-level winds and the surface weather patterns. An example of this association is depicted in figure 9, where the flow pattern of winds at 10,000 feet above the earth's surface is shown in relation to a typical frontal system as shown in figure 8. The effect of the upper-level trough at the 10,000 foot level is analagous to a bulldozer pushing dirt across a field—that is to say, the upper-level trough technically pushes the colder air mass behind the cold front. The surface air rises up ahead of the cold front and over the cold air, much like the dirt along the bulldozer blade. It condenses, forms rain clouds, and results in rain. It is usually true that the deeper and more sharply defined the

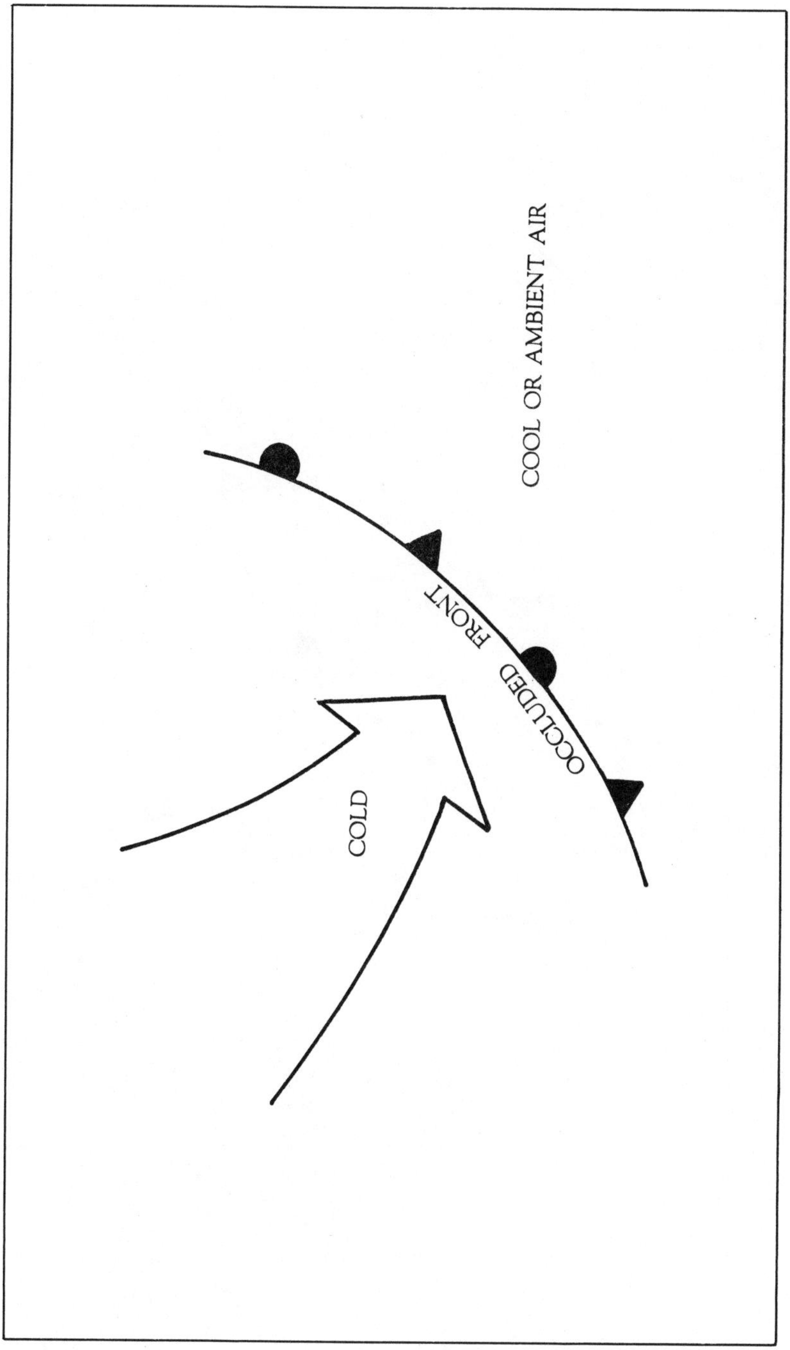

FIGURE 7. OCCLUDED FRONT

COOL OR AMBIENT AIR

OCCLUDED FRONT

COLD

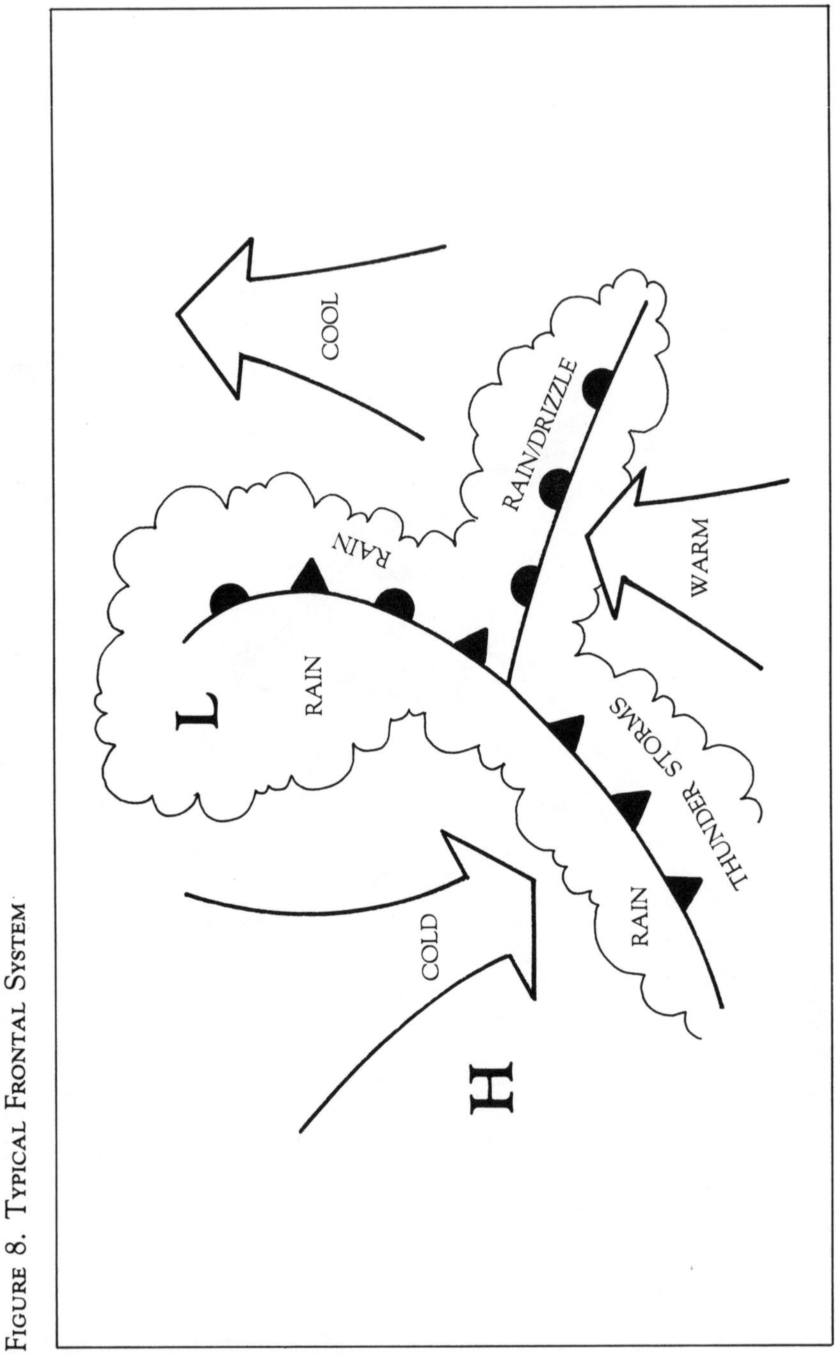

Figure 8. Typical Frontal System

Figure 9. Typical Frontal System at Earth's Surface, Overlaid by Flow Pattern of Upper-Level Winds at 10,000 Feet above Earth's Surface

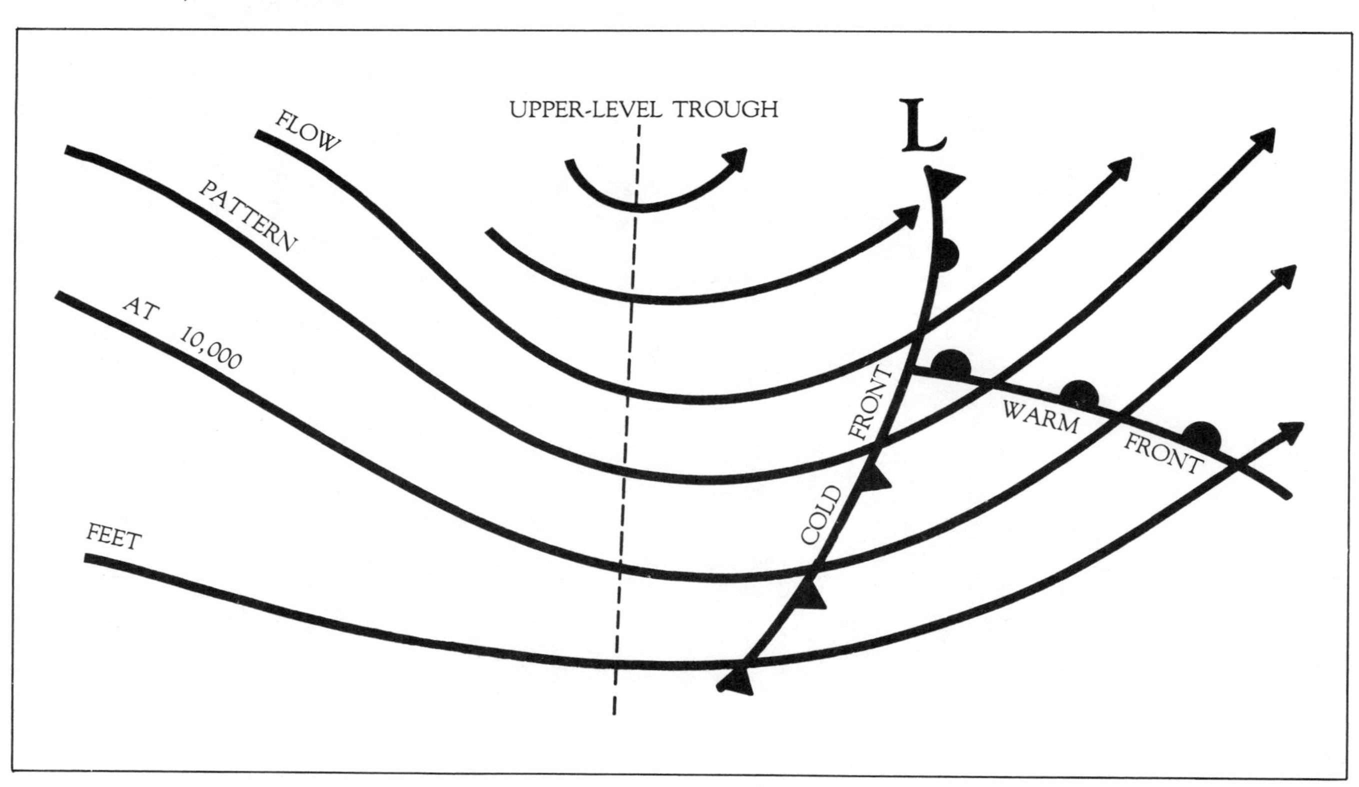

upper-level trough is within the flow pattern of upper-level winds, the greater the intensity of the frontal weather and the more likely it is that precipitation will occur.

THE FINAL STEP: LOCAL FACTORS

You'll remember that we shuffled into the office before dawn, organized our charts and teletype paper, reviewed the upper-level winds, jet stream flow, and surface weather charts. It's about 30 minutes later and we're ready for the final step in our forecasting process. We must now consider all possible local effects, either man-made or natural, which can influence our forecast. There are many factors involved here: topography, sources of moisture (lakes, rivers, oceans), industrialized and developed areas, etc. For example, there will usually be a prevailing wind direction (a direction from which the wind usually blows) for any given location. The direction will be related to the local topography which acts to steer the wind in a certain direction. Topography can also produce unique weather patterns in areas where the land rises very rapidly—as along the east slope of the Rocky Mountains or along coastal regions where mountains are present. As moist air flows from the lower elevations to the elevated land, it will condense because of this rising motion and precipitation.

The Great Lakes play a major role in the weather over the Northeast by fueling the frequent frontal systems in that area with an ample supply of local moisture. The Gulf of Mexico is the major source of moisture for the eastern two-thirds of the nation. Without this moisture it would be very tough to get sufficient precipitation in the major grain-producing areas of the U.S.

Industrialized areas often play a very important, although quite local, role in altering local precipitation patterns. Several studies have shown a very significant increase in precipitation downwind of industrialized areas as a result of pollution, which serves to increase the amount of particles on which rain droplets can form. Cities also induce significant local variations in weather. A very common phenomenon observed in large cities is the "heat island" effect, where the concentration of buildings and paved streets, usually in the downtown area, creates an island of consistently warmer temperatures—often as much as ten degrees warmer than in the suburban areas (see figure 10).

An old saying among salesmen is "you gotta know the territory." This is just as true for the weather forecaster. You've got to be familiar with local effects of weather which consistently appear as a result of any or all of the factors discussed above. Furthermore, it takes at least a year to become familiar with the local effects, because you need to observe the special

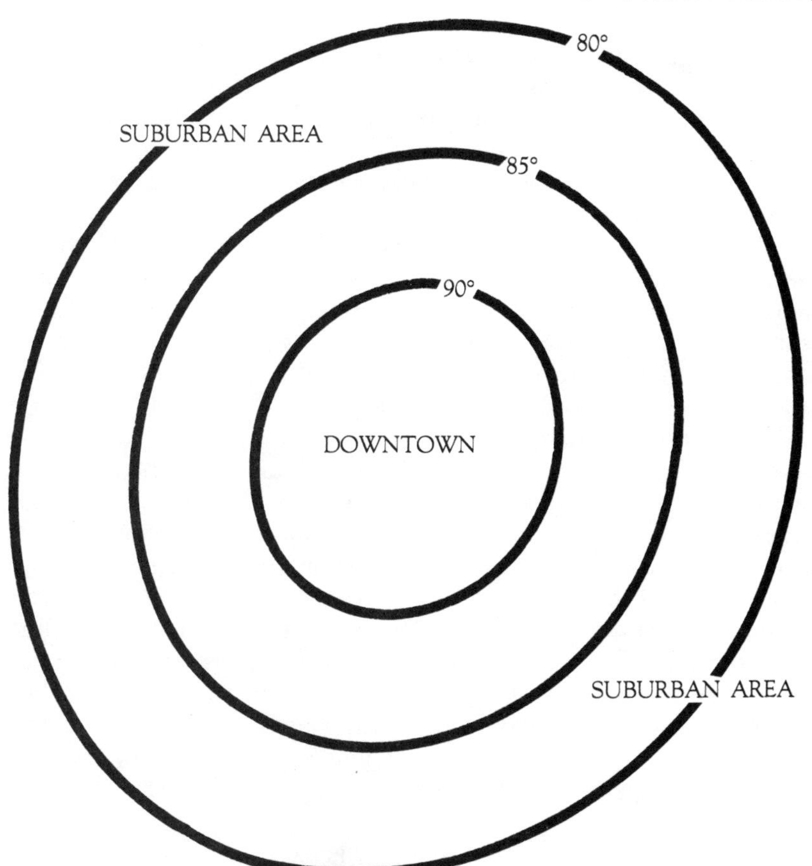

FIGURE 10. HEAT ISLAND EFFECT

differences in these effects in each season of the year. This is a key ingredient in effective forecasting for any area.

PUTTING IT ALL TOGETHER

You are finally ready to put it all together. The entire process has taken less than an hour but the amount of information you've covered is considerable. There is room for error. There are often aggravating gaps in crucial data essential to your forecast preparation. However, when all systems are operating smoothly and you have an experienced forecaster at the controls, the results are usually quite pleasing indeed. The finished product will take various forms, depending upon which special user is to receive a forecast, but the most common avenue of communication is radio, television, or telephone. A forecast should include the amount of cloud cover (an aviation forecast will also include the level of each cloud layer), the amount, intensity, and type of precipitation, and the pressure, temperature, dewpoint, wind direction and speed, plus any special remarks or amplifications necessary to complete the forecast. The terminology and communication of a forecast will be discussed in greater detail in chapter two.

Forecasting is obviously a highly complex science requiring years of training and practical experience. I often compare the typical weather station environment to an emergency room in a hospital. When the weather is bad, the forecaster (like the doctor) is besieged with requests for information while trying to perform several different functions in order to keep up with the rapidly changing weather.

The actual process of forecasting described may have surprised you because of the emphasis on upper-air analysis. This is not a new technique but the emphasis and the technique vary from forecaster to forecaster. However, the dominance of the upper-level winds on the general weather pattern is accepted by virtually all professional forecasters. Upper-level analysis is still a relatively new tool for forecasters and there is much room for improvement. As our use of upper-level data becomes more sophisticated, the overall accuracy of forecasts will increase proportionately.

I hate to use statistics to prove a point, because they're so often misused, but you need something with which to gauge our average performance. The following table will give you a feel for how accurate a professional, experienced forecaster is supposed to be for any given forecast:

TABLE 1. APPROXIMATE FORECAST ACCURACY

LENGTH OF FORECAST	% ACCURACY
2-Day	90%
3- to 5-Day	80%
6- to 10-Day	70%
30-Day Outlook	75% (Temperature)
	60% (Precipitation)
Seasonal Outlook	60-80% (Temperature)
Seasonal Outlook	60% (Precipitation)

These statistics will not hold up for every day of the year but you should expect a real "pro" to come very close to the accuracy values listed above.

Our science is highly developed and, for the most part, we're quite good at what we do, but I'll admit there have been times when I've wished for a trick knee, a coin to flip, or a high quality dartboard! But, all in all, we're providing a valuable service to society and, once we've improved our ability to communicate our service, we'll be even more useful.

2
COMMUNICATING WEATHER INFORMATION

Back in the early 1880s, a frost warning for the tobacco crop near Madison, Wisconsin, was telegraphed nearly 36 hours in advance. However, a local telegraph operator didn't relay the message and a killing frost occurred without warning. The crop could have been saved if the farmers had been notified in time.

Regretably, nearly 100 years later, farmers are still frequent victims of communication gaps. How, in this age of constant communication and a vastly improved weather network, can the public still not be assured of receiving critical weather information?

In 1878 the cold-wave flag was introduced to several major cities and was eventually in use in 290 towns and cities by 1886. The flag, which measured six by eight feet with a two-foot square black center surrounded by a white border, was used to provide as much as 30 to 36 hours advance notice of a significant drop in temperatures within a single 24-hour period. This system worked well for its time and was well received by the public. For example, an editorial in 1876, in the *New York Tribune,* praised the National Weather Service and stated that even though forecasts were not always correct, they were definitely an asset in making plans for the day. Hence, in the late 1800s weather forecasts were already available to most people on a reasonable timely basis via telegraph, newspaper, and a network of weather flags. Furthermore, there was little confusion about the meaning of the forecast—the flags didn't speak and could only mean one thing!

Today we are literally inundated with weather information via radio, television, newspapers, magazines, and in many locations the new federal government weather radio network. Rather than improving the situation, though, the modern day wealth of weather information derived from these sources seems to be more complex and more confusing than ever. Excellent weather information is available to nearly everyone but you've got to know what you're looking for and where to find it.

The ability and amount of knowledge possessed by the modern meteorologist is formidable, but we're encountering problems with communicating our product to the public. There are tremendous problems in insuring that the very latest information is being distributed on radio and television. There are also major problems with getting both the public and special users simply to understand what we're trying to say.

For example, do you know what "50% chance of showers" means? In one rather infamous survey, it was discovered that as many as nine out of ten professional meteorologists didn't know what it meant. It's supposed to mean that there definitely will be showers but that they'll occur over only one-half or 50% of the area for which the forecast is issued. Therefore, it is supposed to refer to the area and not the probability or actual chance for showers. But virtually everyone believes the latter, including most meteorologists.

If a farmer applies several thousand dollars worth of chemicals when the forecast calls for a 50% chance of showers, it is because he is assuming that there is only a 50/50 chance for rain on his land. However, he should have assumed that there definitely will be rain and could even be a 100% chance for rain on his farm if his land is in a rain track. (Precipitation usually falls in the same path or track for as long as a growing season, although this path will vary from year to year.) So, depending upon your land and its relationship to local rain patterns, you need to start getting concerned or elated (whichever the case may be) when the forecast calls for a 30% or greater chance of rain, because there's an excellent chance for rain over at least 30% of your area.

For a closer look at terminology commonly used by weather forecasters, I refer you to appendix C. You will derive far greater benefit from listening to a weather forecast if you'll become familiar with these terms. However, you will quickly deduce that there is a most confusing array of terms and a considerable amount of variability in their meaning and application. For example, one man's partly cloudy could very easily be another man's mostly cloudy or another's mostly sunny. If you're confused by the sky or cloud-cover terms, just remember that mostly sunny is more sunny than partly cloudy which is more sunny than mostly cloudy which is less sunny than partly cloudy!

A further difficulty in using weather terms is faced on days when the weather is changing rapidly. A good forecast should indicate the times when the cloud cover will increase from mostly sunny to partly cloudy or greater amounts of cloud cover, and the amount and intensity of changes of all other applicable weather variables as well as the time of change. An experienced forecaster will be able consistently to predict the timing and intensity of these changes. For example, let's assume that a cold front is approaching your location and that it will be sunny at dawn, mostly sunny by midmorning, partly cloudy by midday, mostly cloudy by mid-afternoon, partly cloudy by dusk, and clear by midnight. At the same time, light rain and thunderstorms will occur over 30% of the local area. Hence, the forecast should probably read: mostly sunny this morning, becoming mostly cloudy this afternoon with a 30% chance of showers. However, it is more commonly stated: partly cloudy today with a 30% chance of showers or increasingly cloudy with a 30% chance of showers.

Ideally, a forecast should contain the major changes and the timing of those changes plus the intensity, duration, and coverage of precipitation. Using this guide, then, the above forecast would become: it will be mostly sunny this morning becoming partly cloudy by midday and mostly cloudy this afternoon with rain and thunderstorms over 30% of the local area (one-half inch average rainfall where the showers occur), showers ending by early evening with clear skies by midnight. I, personally, use this type of forecast terminology because I feel that it is far more informative to the listener once he becomes accustomed to the subtle differences in my style.

So, in order to get the full meaning from a weather forecast, you must be aware of the individual forecaster's personal style. Learning the meaning of the terms in appendix C will aid you appreciably, but you will need to keep in mind that not all forecasters apply the same meaning to each of the terms. There is a considerable amount of latitude in the application of weather forecast terms by forecasters.

SOURCES OF WEATHER INFORMATION

Once you've learned how to interpret the meaning of a forecast you must then seek out the best sources for this information in your community. This may prove to be a greater task than you realize. There are essentially no nationwide standards or forms of control to insure that only the most professional weather information is being used by the respective communications medium—it's usually anything goes just so long as it looks or sounds good. However, the blame lies with both the producers and the users. We professional meteorologists, the producers, lack any meaningful industry-wide standards. (More about this in chapter three.) Radio, television, newspapers, and magazines are also to blame for not making much of an effort to provide the highest quality product.

Radio and television are undoubtedly the two most widely-used sources of weather information in a community. For sheer impact on the community, I give my vote to television. Almost everyone in a community will know the name or recognize at least one of the local TV weatherpersons. Most television executives will admit that weather is undoubtedly the biggest draw in the news program. I am often reminded of this when my farming customers will ask for my opinion the next morning about what their local weatherperson had to say about the weather the previous evening.

The problem with television weather programs is that the trend over the last couple of decades has been away from a quality, professional product and toward an ever-increasing emphasis on eye and ear appeal and assorted gimmickery or buffoonery. Far too many television weatherpersons possess no knowledge whatsoever of weather but do possess a beautiful body and/or melodious voice or have a gift for the comical. It is just as bad to see a totally unappealing television personality who merely presents whatever forecast happens to be available at the time of the telecast. In my travels across the nation and in my discussions with people from other areas, it is obvious to me as a professional that television executives get very low marks for the quality of their weather programs. This might not upset many people, but considering the ominous impact of television on society and the very real need for only the best weather information available, something really needs to be done to improve the situation.

What can be done? The answer lies with a combined program of action by commercial meteorologists and you the viewer. We commercial meteorologists must convince television executives that better weather information exists, the community needs the best weather information available, and that they'll make more money by producing a quality weather program. They'll soon discover that they can keep the eye-appeal of their programs, but by increasing the quality they'll also serve the needs of the community and attract more devoted listeners who appreciate that little touch of quality so rarely seen in the gaudy world of modern television.

You, the viewer, can speed up the process by calling or writing your local television stations and voicing your opinion. Television executives usually rely on sophisticated statistical surveys to gauge the success of their program format, but they also consider the comments of those few viewers who take the time to offer their opinions. I encourage farmers and farm organizations to call or write their local television stations and ask them to provide the best possible weather information. Farm organizations in conjunction with chemical and implement advertisers can bring a considerable amount of economic pressure on any television station to provide better agribusiness and weather information. You may have more power over the television stations than you realize.

HOW TO GET BETTER WEATHER INFORMATION

Let's suppose that you do take the initiative. What do you ask for? The best possible situation is to have an experienced, handsome, and articulate professional meteorologist to give the weather telecasts on at least one local station. But meteorologists of such description and ability don't exactly grow on trees. The next best situation is to have a pleasing TV personality, with a reasonable knowledge of weather, supported by a local commercial meteorologist who can work personally with the television weatherperson to insure that very little is lost in the translation from professional to broadcaster to you. This too, though, is not likely to be all that common an arrangement in the near future because of the lack of commercial meteorologists in all but the largest cities. There are several commercial meteorologists attempting to forecast for cities as much as a thousand miles or more away. It is difficult enough to issue precise forecasts for a community less than 100 miles from your location, much less several hundred or thousand miles away. You will always be more accurate in your own location and it is rare indeed that a forecaster a considerable distance from your location can match your accuracy. Using a commercial meteorologist more than 100 miles from your location should be considered very carefully and depend on that forecaster's proven ability to outdo any local sources.

The most common situation is to have a professional broadcaster, completely untrained in meteorology, who must rely on weather teletype and map data from the National Weather Service (NWS). This can be relatively effective if the television announcer will take the time to read a few books about weather and attempt to understand all of the information available to him via teletype and weather map machines. In major cities the announcer will also have the convenience of an NWS forecast facility. The smart announcer will get to know the government forecasters and become familiar with the forecast facility. This can be a reasonably successful approach depending on the ability and initiative of the announcer.

The potential for weather programs on television is enormous and should eventually prove to be of considerable value to the community. Very little is being done now, and what is being done is rarely of any value to agriculture. Your television station should feature a radar, long-range forecast, and forecasts of greater detail emphasizing variables of particular use in agriculture. An early morning weather program on at least one local station should be mandatory. Forecasts and the timing of the programs are almost always in favor of city dwellers even though the most devoted television weather viewers are undoubtedly farmers.

Much the same situation exists in radio. Radio executives are more or less ignorant of the abilities of professional meteorologists or the marketability of their product. However, the immediate potential for radio appears to be greater than television. The overall impact of television may be greater than radio but the usability of weather information on radio is far greater and more flexible. It is awfully difficult to carry a television with you wherever you go and it is rarely convenient to interrupt a program to give an updated weather forecast unless it involves severe weather. But on radio there are plenty of breaks between songs and there's no problem at all in updating a forecast.

Radio has changed over the past few decades and particularly so in just the last couple of years with the emphasis on FM radio. FM radio allows listeners to enjoy their favorite music with a minimum of interruption. News and weather generally do not play a major role on FM radio. Consequently, AM radio has become more news and information oriented. The emphasis on AM is on the personality of the announcers and their ability to communicate with the listener, and on the ability of the news department to provide the very latest developments in news and weather.

Most farmers would agree that agricultural news and weather information via radio is one of their most vital sources of information. There are few places that a farmer goes where he can't take a radio along. He may not always like the music or the announcers, but if he needs and uses the information, he'll listen. Therefore, I would encourage farmers to concentrate their efforts initially on AM radio to insure that at least one local AM radio station is providing the best possible and most frequent weather information (no less than every 15 minutes with a full five-day forecast each hour).

The guidelines for upgrading weather information on radio are the same as for television. A professional meteorologist experienced with local weather patterns is always the best bet. Here, the emphasis is almost totally on the quality of the voice with little or no regard for appearance. Meteorologists with good voices are quite plentiful; in fact, there is an abundance of retired military weather forecasters, with superb experience, in practically every city of any consequence—a virtually untapped gold mine for agriculture as well as the radio industry.

Here again, you should write or call your local radio station and voice your concern for better weather information. You will probably have to concentrate your efforts on the most powerful AM station in your area to insure that your weather information will be broadcast on the strongest signal with the least possible breaks. Also, the more powerful AM stations will usually be richer and more likely to entertain the idea of spending the extra money necessary to upgrade their weather information to meet the standards you request. Once they've made this expenditure they will undoubtedly be pleased at the return on their investment, for they'll quickly discover that advertisers like to spend their money on weather programs.

If you can't convince the larger AM stations to upgrade their weather information, you might be able to interest a smaller local station in subscribing to the National Weather Service weather teletype network (NOAA Weather Wire). Though this network is quite inexpensive, it provides a wealth of current weather information, radar reports, climatology data, agricultural weather forecasts, and a great deal more. In fact, I encourage all small radio stations to subscribe to the NOAA Weather Wire, for it is undoubtedly the least expensive way to provide quality weather information to their listeners.

IDEAL WEATHER PROGRAMMING

Ideally, at least one local radio station should feature an early morning weather program with the closing commodity prices and a detailed agricultural forecast, plus a look at weather across the major agricultural areas in North America and how this is likely to affect crops in those areas, which might have a bearing on the marketability of crops in your area. The latest weather forecast should be read by the announcer at least every 15 minutes with a minimum five-day forecast at least once an hour, and longer-range outlooks whenever possible. Commodity prices of local interest should also be given once each hour through the day. Radar reports should be given whenever any precipitation is detected within a one-hundred mile radius of the radio station, and severe weather reports should be read as quickly as they are received.

There is vast potential for both the listener and the communications executive in having the best possible weather information on radio and television. The listener, particularly the farmer and other special users, will benefit immeasurably from timely and accurate weather information that's actually usable on a daily basis. The radio or television station will enhance its community reputation and make more money by marketing weather programs which are always a good bet with advertising accounts.

There is another radio network (briefly mentioned earlier) which should be discussed before we move on to written publications. In the 1960's the National Oceanic and Atmospheric Administration (NOAA), the parent agency of the National Weather Service, initiated a network of radio stations broadcasting continuous weather information for boaters.

This network, now called NOAA Weather Radio, has since been expanded to over 200 stations in nearly every state in the union and is due to be completed by 1979 with over 340 stations broadcasting continuous weather information to over 90% of the population.

NOAA Weather Radio is a definite asset to the meteorological community and to the general public. I encourage all of my customers to purchase the special radios required to receive the broadcasts. The broadcasts supply current weather conditions (including radar reports) during significant weather, severe storm warnings (some radios will even alert you automatically), and continuous weather forecasts for your general area. The limitations to this system are that you must have a special radio and the effective radius from each broadcast point is only about 40 miles.

WRITTEN COMMUNICATIONS

The written media can, of course, never compete with the currency of electronic communication, but they do play an important role in weather. Nearly every daily newspaper in America carries at least some mention of the local weather forecast. Most of the larger newspapers print detailed weather summaries and forecasts for at least the local area and usually show a map of the continental U.S. with forecasted weather patterns. The weather information printed in daily newspapers is often helpful but rarely current enough to satisfy most users. I would not encourage using a newspaper forecast to make an operational decision for your farm.

There are many agricultural periodicals, and quite a few feature long-range weather forecasts. The best way to judge these forecasts is, quite simply, to take the time to verify their accuracy for your area. If the forecast is accurate most of the time, use it. I will discuss long-range forecasts and the accuracy you should expect in chapter four; use this as your guideline.

You might also want to check out several weather publications produced by commercial meteorologists. First, compare the credentials of the meteorologist to those given in chapter three. If the meteorologist appears to have the necessary qualifications, you might want to try a subscription to his publication if it appears to be of use to you. Beware of those who promise you more than you can reasonably expect according to the guidelines prescribed in chapter one.

The U.S. government has quite an array of weather publications with several specifically oriented toward agriculture. I recommend that you subscribe to the *Weekly Weather and Crop Bulletin.* The *Bulletin* does an excellent job of keeping you informed about the effect of weather on crops in all major agricultural areas of the U.S., Canada, and most areas of the world. I also recommend that you obtain a copy of the *Local Climatology Data (LCD)* for the nearest Class A observing station of the National Weather Service. (You may call or write the National Climatic Center in Ashville, N.C., to obtain an *LCD.)* The *LCD* will give you an excellent idea of the average climate conditions for your general area and provide a comparison for the climate data you collect for your farm. It would be a good idea to contact the nearest National Weather Service office or commercial meteorologist to inquire about additional publications which might be of interest to you. Nearly all of the government publications are quite inexpensive.

In my talks to civic groups I always conclude by saying that there's plenty of smoke but very little fire in professional meteorology. We're making a good steady flame with a consistently good product, but very few people are aware of it because of the smoke emanating from inefficient communication.

There are problems with television and radio because their executives are generally not aware of the ability of properly trained meteorologists. They already know that weather is a highly marketable product but it will take years to convince them that they can improve the product, better inform the public, provide a needed public service, and at the same time, make more money.

It will also take several years to produce meteorologists properly trained for radio and television communication. Universities will have to initiate better training programs for potential meteorologists. Professional meteorologists of the future should be fully versed in all the various techniques and terminology for effective communication of our highly complex science the moment they graduate from college.

For now, however, you are more or less at the mercy of the communications industry. It is ironic that you have available to you more weather information per radio station or printed page than ever before, yet it is more confusing than ever. Your great-grandfather would know that a cold snap was imminent when he saw the cold wave flag at the town square. However, there are days when you've listened to six different forecasts on radio and television, read yet another in the *Daily Disappointment,* and, quite honestly, don't know whether to wear a raincoat, put on your snowshoes, or wear a light shirt because of the heat!

Usable, quality weather forecasts are available but where do you find them? I have suggested several ways to improve local sources earlier in the chapter but, depending on your success and tenacity, that will rarely be enough to satisfy your needs. You would do well to locate the nearest commercial meteorologist to see if he or she can help.

There aren't many commercial meteorologists who are familiar with the weather needs of farmers but you owe it to yourself to see if there are any competent meteorologists in your area. Our job is to sift all available data and provide usable weather forecasts for our customers. We do our best forecasting within a one-hundred-mile radius and once we become familiar with your farm, we'll be able to provide weather forecasts specifically tailored to your needs. This is the ideal situation but it will be a good five to ten years before commercial meteorologists, familiar with the needs of agribusiness, can be found in all major agricultural areas of the United States and Canada.

For now, you will probably have to be content with the better understanding of weather and weather terminology you'll get from this book and with your efforts to improve local sources of weather information. But there is hope on the horizon.

3
THE ROLE OF THE METEOROLOGIST

In my opinion, the professional meteorologist will play an increasingly important role in society. Within a decade the Certified Consulting Meteorologist (CCM) will be on par with lawyers, accountants, and professional consultants of similar status.

In the future, the CCM will consult with radio, television, industry, agriculture, and individuals. He will provide a respectable product that will keep the general public better informed about the weather and serve the specific needs of special users. The benefit to agriculture alone will be on the order of several billion dollars annually. For example, if every farmer in America understood what 50% chance of showers meant and how that forecast specifically affected his farm, how many hundreds of millions of dollars annually would be saved through more efficient and timely applications of pesticide, herbicide, or fertilizer?

By 1990, CCM's will probably be found in every major city in the U.S. and in most cities with a population of 50,000 or more. They'll be consulting with city, county, and state government, local industry, pilots and aviation concerns, engineering firms, commodity brokers, farmers, etc. They'll be issuing detailed forecasts for special users as well as the general public. You may even pay a CCM a nominal fee to prepare a forecast for a backyard barbecue or your annual vacation.

The CCM of the future will, of course, play a vital role in agriculture. Within a decade it may be just as common for a farmer to consult with his meteorologist as it is for him to confer with his lawyer or accountant. The CCM will work with the farmer to improve his knowledge of the effect of weather on his farm and how he can use that information along with the forecasts provided by the CCM to improve the efficiency of his farming operation and marketing practices.

A CREDIBILITY GAP

At the present, however, there is an enormous credibility gap between public opinion of the ability of the professional meteorologist, and his actual capabilities. This gap is understandable considering the mass confusion of weather information generated by the many communications media mentioned in the preceding chapter.

I have already indicted, tried, and convicted the communications media; let me also speak heresy about the profession. Almost anyone, regardless of his qualifications, can call himself a consulting meteorologist, leaving the door open for assorted crackpots and flim-flam artists who have capitalized on the ignorance of the general public to line their pockets. This unethical approach to the business has left a rather rutted road for those of us attempting to project an ethical and competent image for professional meteorology. It will take time to rectify the situation.

We are having difficulty establishing our credibility because of a general lack of consistent standards within the profession. We are just beginning to think of ourselves as professional consultants with valuable knowledge instead of simple chart readers and forecasters with a take-it-or-leave-it attitude. As our self-image improves, we'll begin to enforce better forecast techniques and a more professional standard of ethics industrywide.

But, we've got a long way to go to produce a true set of professional standards for meteorologists. For example, the status of Certified Consulting Meteorologist (our equivalent to the CPA or CLU) is possessed by only a couple hundred meteorologists in the United States. The ranks of CCM's must expand into the thousands before we can realistically begin to impose stricter standards within the profession and improve our public image.

One reason there aren't very many CCM's around is the general lack of regard for private practice among most meteorologists. There have been meteorologists in private practice since the 1800s but it is still rare to see a young meteorologist take that long and frightening step out into the cold, cruel world.

Industrial or commercial meteorology, though not new, only really got its feet on the ground after World War II. The war resulted in giant strides forward in weather technology and, subsequently, analysis. This sudden improvement in meteorology allowed, for the first time, the industrial meteorologist to make consistently excellent short-range forecasts for special users in industry, government, and agriculture.

However, the development of commercial meteorology has been agonizingly slow for two primary reasons: 1) the massive involvement of the federal government in meteorology, and 2) the inability of most meteorologists to see the fantastic potential in private practice of commercial meteorology. The federal government has, so far, been the only organization large enough to fund a national weather monitoring and analysis network and a sufficient corps of weather forecasters to serve the entire country. This isn't likely to change in the immediate future although, as commercial meteorology continues to develop, the role of the National Weather Service should become more general and more research oriented with less and less application at the local level. The National Weather Service will continue to serve a vital function in the collection, analysis, and transmission of weather data and in setting the basic standards necessary to make such a massive network functional and efficient.

The commercial meteorologist will gradually begin to take up the slack at the local level. The increase in CCM's and improvement in our public image will begin to produce a new wave of professional meteorologists in private practice, who will form the basis for the development of what should quickly become one of the nation's next multibillion dollar service industries.

To meet this challenge we must begin to improve our image by setting stricter standards, promoting professionalism, and providing for more effective training which will allow the young graduate meteorologist to enter private practice sooner and more effectively. The few universities that offer a degree in meteorology are generally turning out educated idiots, with no practical experience, who must first be trained by the military or government before they're able to enter private practice. We should begin now to offer a training program for potential meteorologists which will allow them to enter private practice immediately out of college or at least prepare them for employment by a commercial meteorologist upon graduation. The budding prognosticator should be given healthy doses of experience in radio and television broadcasting, effective public speaking, statistics, computer programming, and the very latest weather forecasting and analysis techniques, in addition to the standard fare of meteorological theory. The young meteorologist should be well versed in the basics but also able to communicate his product to his clients effectively.

HOW TO SELECT A METEOROLOGIST

By now, you can see that we've got problems in professional meteorology but the future looks bright and should prove more beneficial to us all. However, I've not yet shown you how to select a meteorologist who can do the job for you.

What you need is either two separate people or the rare individual who combines both the ability to forecast the weather and the broader knowledge of application of weather information to your special need. Experienced forecasters, though not exactly growing on trees, are in relative abundance. The best forecasters available are usually retired Air Force sergeants with ten or more years of practical experience in weather forecasting in several locations across the nation and the world. In my own experience as an employer, I have generally found these individuals to be superior forecasters in the two-to-three-day range and, with a little instruction, in long-range forecasts beyond the five-day range. There are undoubtedly a couple of these individuals near you who could be persuaded to consult with local radio and television and provide you with at least one credible source of weather information in your area.

Best of all, of course, is the person who not only knows how to forecast but also knows how to apply weather data to your use. This individual will usually be a professional consulting meteorologist with at least one advanced degree and several years of experience in both forecasting and special applications of weather information. The consulting meteorologist should be able to guide the farmer through the procedures outlined in chapter five and suggest special applications of these procedures to your individual farm. This person is a rare breed indeed; there are currently only a handful of them nationwide. Their numbers will increase slowly over the next decade.

In conclusion, there are many people who call themselves professional meteorologists, so it's not easy to pick the real "pro." You can narrow the odds by first considering their experience. The more experience a forecaster has, the better he usually is. A degree in meteorology does not imply forecasting ability; it still takes at least three years after graduation to become a respectable forecaster.

A consulting meteorologist will generally possess several years of forecasting experience as well as several years of practical application of his knowledge in a special area. He should have at least one degree, preferably in meteorology, and be a member of the American Meteorological Society. It is also a plus if he is a Certified Consulting Meteorologist.

A farmer owes it to himself to seek out the qualified professional. Weather plays such an important role in farming that it can't be ignored—as it largely is today.

4 CLIMATE VARIABILITY AND LONG-RANGE WEATHER FORECASTING

Several years ago prehistoric animals were found perfectly preserved in the frozen wasteland of the Arctic. In the digestive system of these animals there were still particles of recently eaten vegetation commonly found in warmer climates.

What could possibly have happened that could have deep-frozen the animals instantaneously and how could they have been eating vegetation not found in the arctic zone?

One proposed explanation of this phenomenon came from the Continental Drift Theory. The hypothesis was that a sudden catastrophic shift of the earth's crust instantaneously displaced the continent on which the animals existed from a subtropical latitude to the arctic zone.

Most climatic changes are not nearly as violent, but there have been many extreme changes in climate since the formation of the earth. Geological records show four such climatic reversals over the past three to four million years—i.e., the Pleistocene epoch. Each major change was characterized by massive ice sheets which spread southward. The five glacials were coincident with pluvials further south which brought rain to the desert areas of the world.

During the peak of the ice ages, as much as nine percent of the earth's surface, or thirty percent of the land surface, was covered with ice. This compares to less than four percent of the earth's surface covered with ice today. There was less rainfall worldwide during the glacial ages. The major rain tracks were shifted further southward to where the great deserts are now. Precipitation was two to three times greater in these areas than today. The advance of the glaciers resulted in herds of walrus off the coast of Georgia, spruce forests in Texas and Florida, and reindeer in the central United States.

The average temperature during glacials was about 5 to 6 degrees Fahrenheit below today's. It was 3 to 4 degrees Fahrenheit above today's during the interglacials (periods between glacials typified by warm, dry, ice-free climates). Today's climate is nearer interglacial than glacial but leaning in the direction of yet another glacial.

The above discussion may seem irrelevant but the point is that climate is just as variable as the daily weather—it just takes longer! Also, the implications are far more drastic. A hurricane can devastate a local area; a global change in climate can have an impact on every household in the world.

Changes in the global climate are always evident but the differences are usually only considerable when viewed over a century or more. For example, before the 15th century, grapes were widely cultivated in England and the French complained of the impact of English wine on European markets. As early as the 10th century the Vikings had established prosperous colonies in Greenland, having named the island for its lush green pastures. By the early 15th century, however, the Viking colonies were wiped out by cold and hunger, and now four-fifths of Greenland lies buried under hundreds of feet of ice. The climate in England is now much too harsh to support the vineyards of old.

Much quicker changes can occur. For example, in April 1815 the volcano Tambora erupted in the Dutch East Indies. This was the largest known volcanic eruption in recorded history. From 1811 through 1818, an estimated 220 million metric tons of ash were injected into the stratosphere as a result of volcanic activity. Of this, 150 million metric tons have been attributed to Tambora alone. The volcanic ash spread worldwide in a few weeks and did not sink out of the atmosphere, totally, for several years. During this time, the ash reflected a considerable percentage of solar radiation back into space. The atmosphere cooled. The result was a worldwide decrease in temperature in 1816 when there were frosts in Pennsylvania every month of the year, a circumstance altogether without example. Crop yields were cut drastically worldwide in an extremely short growing season. The effect on the price of flour, for example, was devastating—it more than doubled by June of 1817.

It is interesting to note that we may be living in double jeopardy with respect to the injection of particles into the stratosphere which are large enough to reflect solar radiation. We can normally expect an average of five very large volcanoes to erupt each century. However, the modern period of great technological activity has been free from major volcanic eruptions to an almost historically unique extent even though the amount of volcanic activity worldwide is on the increase. At the same time, at the supposed current rate of man-caused injection of particles into the stratosphere, the density of these particles would be near the levels produced by Tambora by the year 2039. One can only hope that in the meantime a volcanic eruption of the order of Tambora does not further aggravate the situation.

The global climate is in a most delicate and, geologically speaking, unstable balance. We are deluding ourselves if we think otherwise. In fact, the current trend is quite possibly in the direction of yet another ice age, the first major effects of which could be witnessed within the next fifty to one-hundred years.

For nearly half of the current century, mankind was blessed with one of the most benign climates of any period in at least a thousand years. During this beneficent era the human population more than doubled. Statistically, we have about a one in ten-thousand probability of duplicating such weather over the latter half of the century.

There is now good reason to believe that the world climate is rapidly reverting to a less beneficent pattern of weather. A pronounced warming trend was evidenced between 1890 and 1945 but, since 1945, the global mean temperature has been dropping. The effects in some parts of the world have been substantial. Icelandic fishing fleets that used to range northward during the warm period have now had to return to traditional waters to the south. For the first time in this century, ships making for Iceland's ports have found navigation impeded by drifting ice. Since the late fifties, Iceland's per-acre yield has dropped 25 percent.

In North America the armadillo extended its range as far north as Nebraska during the warming trend but is now retreating southward. In England the average growing season is two weeks shorter than it was prior to 1950. Global temperatures since 1945 have undergone the longest unbroken trend downward in hundreds of years.

There are many, many theories which seek to explain glaciation and the reasons for the glacials of the Pleistocene epoch. All of the theories try to determine the specific cause for glaciation. This may well be a futile effort. Whatever the actual cause, the most important ingredient in determining the earth's climate is, of course, the sun.

Temperature on earth is determined primarily by the radiation balance. About 30 percent of the energy radiated from the sun that strikes the earth is reflected back to space. Twenty percent is absorbed by the atmosphere and the remaining 50 percent is soaked up by the land and the water. The 70 percent absorbed by the earth and its atmosphere must be balanced by an equal loss re-radiated back to space in the form of infrared radiation, or an imbalance will occur allowing the earth and its atmosphere to cool off or heat up. There has been no shortage of theories as to why this radiation balance might be tipped so that the earth gains or loses heat at some times more than others.

Generally, there appear to be two schools of thought concerning climate change which can be loosely referred to as the natural and man-made schools. The former school ascribes to the idea that natural causes determine the long- and short-term trends of climate and that man has little or no influence. Natural causes would include periodic variability in solar radiation which can result in a relatively rapid 5 percent drop in total solar radiation—enough to trigger the sequence of ice ages that have repeatedly frozen the earth over the past millions of years. Another natural cause would be an increase in the mean annual snow and ice-pack coverage which could reduce atmospheric temperatures and encourage the subsequent accumulation of more snow over a wider area and the eventual formation of glaciers.

On the other hand, the man-made school suggests that, despite the continuation of natural climatic fluctuations, human activities will be the main contributor to future climate changes. It is argued that man is increasing the amount of particles and carbon dioxide in the atmosphere. The combined effect is likely to be one of warming the atmosphere by trapping solar radiation, as in a greenhouse, and tilting the radiation balance in the warm direction.

For the short term, the abundance of carbon dioxide resulting from man's activities has supposedly kept the temperature from plummeting even further than it already has since 1945. But the trend, for the present at least, appears to be toward cooler temperatures largely because of a steady increase of particulate matter within the upper atmosphere.

Even though a trend towards colder weather exists, we do not now possess the scientific ability to confirm this trend definitively. But it should be more than obvious to you by now that the world climate is most variable. If we acknowledge this variability and our vulnerability to it, we can do some important long-range planning.

For instance, the major grain-producing areas of the world are primarily north of 40° north latitude and hence, quite susceptible to very harsh weather patterns at the critical stages of growth for wheat, soybeans, and corn. The major grain-producing regions of Canada and Russia are particularly vulnerable since they are primarily north of 50° north latitude. A sudden, drastic change in the world climate (as with a large volcanic eruption) could result in extremely damaging cold and wet weather around the globe, at the critical stages of plant growth and a subsequent widespread crop failure. Our grain reserve would then prove to be easily as valuable to national and international political and military stability as our more publicized, yet inedible, nuclear arsenal. Armed with this knowledge, a mandatory program for establishing grain reserves sufficient to sustain us through a massive worldwide crop failure would seem the prudent course. Climate variability should not be considered as just an interesting scientific phenomenon, but rather, an accepted fact and essential ingredient in the planning and maintenance of our national security.

Considering the drastic implications of significant changes in worldwide climate, we should make every effort to develop methods for long-range forecasting. It may seem like an impossible task but the problem may not be as insurmountable as you might suppose.

LONG-RANGE FORECASTING

Technically, a long-range forecast involves any forecast greater in length than five days. It is quite possible to issue forecasts consistently and accurately in the five- to ten-day range. The technique is similar to that described in the first chapter, although the longer the forecast, the more you must rely on long-term statistical trends.

Generally, the weather likely to occur in your location for the next two to three weeks can usually be inferred from the shape of the jet stream across the continental U.S., out over the North Pacific, and across Siberia. What you are essentially trying to determine is when the parcel of air in one section of the jet stream will be overhead. When this parcel of air is overhead it should result in a specific type of weather. In figure 11 the weather most likely to occur at the location marked by the asterisk is indicated along the jet stream in the position coincident with a particular parcel of air. If you lived at this location (North Carolina), you could very easily determine the weather for the next five days merely by tracking the jet stream out over the North Pacific and analyzing the parcels of air most likely to be overhead on the days indicated. The showers would have reached North Carolina by Tuesday. Wednesday would be partly cloudy. On Thursday, Friday, and Saturday the parcels of air would be coming off the back side of the high pressure ridge and would dry out as they sank in the atmosphere resulting in cloud-free or sunny skies. By the time Sunday arrived, thin wispy traces of high or cirrus clouds would be drifting overhead.

This technique is deceptively easy. It is an extremely complex task and one that can only be performed by a very experienced meteorologist. The key ingredients for success are the ability to determine the speed of the

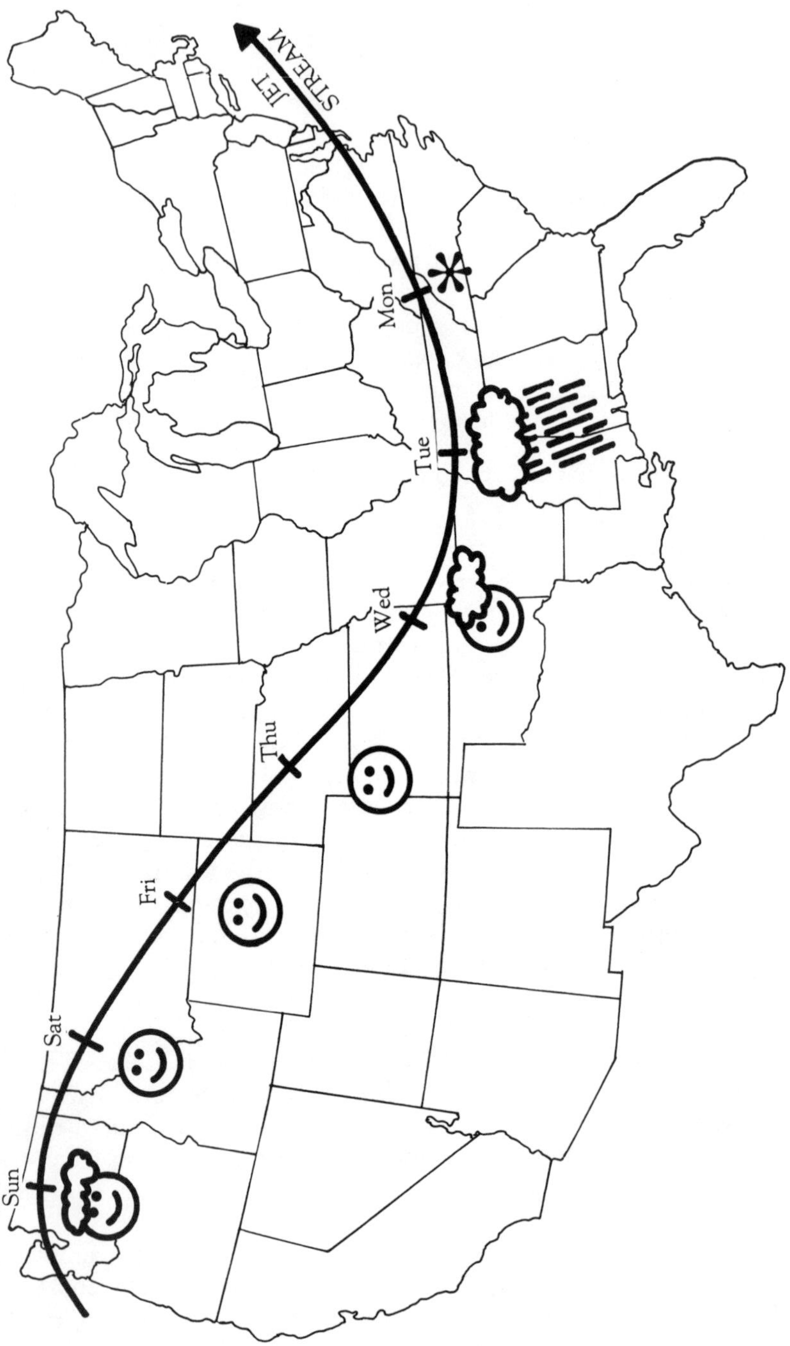

FIGURE 11. LONG-RANGE FORECASTING BY PROJECTION OF PARCELS OF AIR ALONG EXPECTED JET STREAM PATTERN.

individual parcels of air and the future shape of the ridges and troughs and, consequently, their effect on the movement of the various parcels.

You are still not assured of an accurate forecast even if you correctly accomplish all of the above. You must consider the effect of various factors peculiar to your location or related to changes in the composition of the parcel of air before it arrives overhead. For example, a particular parcel of air could result in cloudy skies with no rain over Nebraska, yet result in thunderstorms over Arkansas. Moist air from the Gulf of Mexico would have added enough water vapor in the atmosphere as the parcel of air was in transit from Nebraska to Arkansas to result in showers.

For a forecast beyond three weeks, you must rely on statistical projections based on long-term cyclical patterns of weather and their association with the current trend. To a pilot this is somewhat like saying that you can fly VFR (by visual flight rules) out to three weeks but must fly IFR (by intrument flight rules), or completely by instruments, when you get past the three-week point. Using statistical projections can never be as accurate as tracking the actual parcels of air, but it can provide some useful results.

Long-range forecasting is risky business. Mistakes are bound to happen with either the parcel or the statistical method. By comparison, short-range forecasts of two to three days in length are almost always more accurate and precise than forecasts beyond five days. Just because the risks are greater doesn't mean we shouldn't attempt long-range forecasts. However, because of the greater error involved in long-range forecasts the layman just naturally assumes that the forecaster rarely knows what he's talking about. This credibility gap is regretable but it should not deter the professional from continuing to issue long-range forecasts.

In actual application, there is no fixed point at which the parcel method ends and the statistical method begins. In fact, when the upper-level winds or jet stream become hard to read you may need to rely on the statistical projection to help you prepare your forecast even if it's only five to ten days in length. The forecaster must also not forget that it is usually colder on November 30th than on November 1st and warmer on May 30th than on May 1st! It doesn't pay for us to get so engrossed in our sophisticated techniques that we can't see the forest for the trees.

THE STATISTICAL APPROACH

I have referred to statistical projections several times but haven't yet shown you how they are compiled. How do we determine if a trend exists? Are all statistical trend methods the same? What is the scientific basis for statistical projections of weather? How accurate are they? Can they be used by the decision maker?

Essentially, a trend always exists, for the weather is always in a state of transition either within a season or from one season to the next. This continuous seasonal trend can be considered as the starting or reference point for all long-range forecasts. This basic trend is usually referred to as the normal pattern of weather through a twelve-month period. The long-range forecast should indicate how much, if any, difference there will be in the actual weather versus the expected normal weather. In other words, will November be a typical November or will it be wetter or drier or warmer or colder than normal?

By standard practice the climate or normal pattern of weather for a given location is usually considered as the 30-year average of weather for each month of the year. For example, the average precipitation for July would be calculated by adding the total amount of rain received for July of 30 consecutive years and dividing by 30 to get the average or normal. The same procedure would be used for high and low temperatures.

The 30-year average works reasonably well and, in most cases, provides at least a good first guess for the expected weather in any given month. But because there is always a trend in one direction or another, the 30-year average can never be completely correct for every month of the year. Hence, the need for some method to anticipate or predict when these differences from normal or the 30-year average are most likely to occur.

There is no universal method for preparing statistical projections. There are several sources for these projections, including the National Weather Service. Some sources make exaggerated claims of accuracy, but the truth is that no completely accurate or fail-safe method exists even though consistently excellent results are obtained by at least several reputable sources.

Most, if not all, current statistical projection methods are based on the scientific fact that weather characteristically displays a more or less uniform cycle of eleven years between extremes of hot and cold or wet and dry. This cycle is, in turn, directly related to the eleven-year solar cycle or period of time between a maximum and minimum occurrence of sunspots. As explained earlier, the sun is the most important ingredient in determining the weather. Hence, any variability in the sun's radiation caused by sunspots is likely to have at least some effect on the earth's weather.

A good example of the cyclical nature of weather is the occurrence of drought in the Midwest. John Steinbeck's novel, *The Grapes of Wrath,* vividly portrays the dustbowl days of Oklahoma in the 1930s. Similar, though less intense, drought reoccurred in Oklahoma and major agricultural areas in the Midwest in the 1950s and 1970s. It is only natural to assume yet another round of drought conditions in the 1990s.

The solar cycle is relatively uniform and consistent. A marked increase in the number of sunspots (Solar Maximum) detected within the sun's atmosphere is usually observed every eleven years. This decreases to a minimum number of sunspots (Solar Minimum) some five and one-half years later.

But the solar cycle has not been entirely uniform. For example, there was an unusually long period of minimal sunspot activity in the 17th century. This coincided with a period of much colder-than-normal weather in the Northern Hemisphere.

On an average basis, it is usually safe to assume that the eleven-year solar cycle will be reflected in the weather pattern for a particular location. However, the weather cycle will not usually display the same smooth and consistent variability as the solar cycle. This is easily seen when the actual average temperature for a particular month is plotted versus the normal (30-year average) for a 30-year period. A cycle appears to exist although it is by no means a smooth trend. By compiling a successive series of five-year averages, a much smoother cycle becomes evident. The latter is consistent enough to begin to identify a definite cycle of warmer- or cooler-than-normal temperatures.

This process of smoothing out the erratic fluctuation of actual weather is usually the first step in constructing a long-range forecast model for any given month. The actual model would involve several additional calculations and analyses. Despite a more sophisticated analysis, though, the basic fact remains that no current technique is capable of forecasting the infrequent, yet persistent, erratic fluctuations in actual temperatures. For this reason the actual weather that occurs *can* be completely the opposite of the long-range forecast determined by statistical projection.

A plot of average monthly precipitation for November displays the same characteristics as temperature. A cycle appears to exist but it's difficult to see until the curve is smoothed by calculating a succession of five-year averages through the 30-year period. Precipitation is generally more erratic and the cycles are less definable than with temperature. But cycles definitely exist and an analysis of these cycles will allow you to attempt statistically to project expected rainfall.

What your statistical projection is actually trying to do is to accurately assess the average pattern of the jet stream for the month or period in question. The importance of the average jet stream pattern is that it determines what types of air will invade the continental U.S. Each type of air, or air mass, is named after its source region and each air mass will generally result in a particular type of weather pattern. In winter, cold air masses come from four primary areas: polar region, North Pacific and

North Atlantic, and northern Canada. The resultant air masses are arctic, maritime polar, and continental polar, respectively. The coldest air masses are arctic; the next coldest are continental polar. Maritime polar air masses are less cold and usually wetter.

In summer, arctic air masses rarely move south of the Canadian border. Continental polar and maritime polar air masses invade the lower 48 states in summer but they are much milder and weaker than in winter. In addition, two other air masses play a major role in the summer: continental tropical and maritime tropical. The latter air masses originate over the western Atlantic, Gulf of Mexico, and eastern Pacific. These air masses result in hot, humid weather during the summer and sudden northward surges of unusually warm and humid air in the winter. Continental tropical air masses originate over northern Mexico and result in hot, dry weather in the Southwest and Rockies.

The problem with the statistical outlook is that it can imply two similar jet stream patterns, each of which can result in cold temperatures. But one pattern will produce dry, cold weather and the other—wet, cold weather. The precipitation statistics will, of course, help you in determining whether you'll have wet or dry cold weather, but in actual practice they are just not that reliable for a single month at a time.

Precipitation does reflect a definite cycle in most locations, but there are so many additional factors involved in the determination of precipitation that it is more difficult to call the shots as correctly for each and every month than it is for temperatures.

In addition to cycles of monthly temperature and precipitation, there appear to be definite cyclical patterns in the weather on much shorter scales. Assuming that a cold, wet month is forecast, it is helpful to know which portions of that month are likely to be the coldest and wettest. This can be determined by a combination of factors which include a long-term analysis of the weather that actually occurred on every day of the month for a period of several decades. This analysis of the so-called daily climate can yield some very interesting results.

A sample of a daily climatology for the month of November in Memphis, Tennessee, is depicted in figure 12. If you've already forecasted a cold, wet November, figure 12 could help you to estimate when the coldest and wettest weather would be most likely to occur during the month. For example, whenever a cold front passes Memphis in November it is almost always preceded by warmer air flowing up from the south and followed by colder air moving in behind the front. This phenomenon is most obvious in figure 12 around the middle of the month, which shows a sharp peak in the average daily maximum and minimum temperatures. This also coincides with a day most likely to receive rain, i.e., the 16th.

FIGURE 12. DAILY CLIMATOLOGY, MEMPHIS, TENNESSEE, NOVEMBER

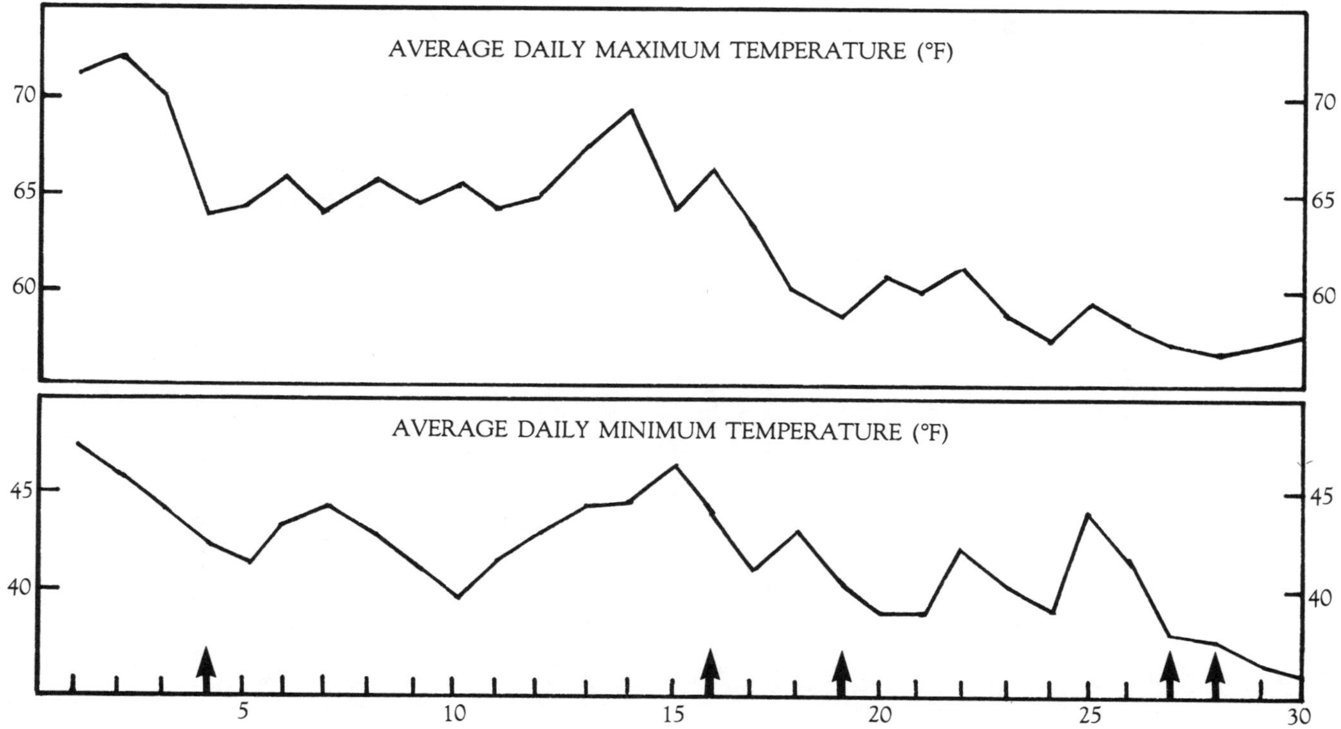

This diagram depicts the average daily maximum and average daily minimum temperature as computed for a 30-year period in Memphis, Tennessee. The days on which rain occurred most often are designated by an upright arrow (⌂).

A cold front usually produces precipitation as it moves through an area. This, too, is implied in figure 12 where the days that received rain most often in the 30-year period are indicated.

Not every month in any location will display a temperature or precipitation pattern as in figure 12. But a pattern of this nature occurs for most months in virtually all of the major agricultural areas, and, when used in conjunction with the statistical projection, a knowledge of this pattern can prove most valuable in determining the specific dates cold and wet weather is most *likely* to occur.

The daily climate analysis can be a powerful tool in long-range forecasting. It can also be misused if you assume that you can accurately forecast the weather for a single day for weeks or months in advance. You can be more accurate than a coin toss but not consistently accurate because you can't determine how well the daily climate analysis will fit the current or actual pattern until that month begins.

THE DYNAMIC APPROACH

So far, the discussion on long-range forecasting has dealt almost exclusively with the statistical approach. There is another side to the coin: what we meteorologists call the dynamic approach—i.e., trying to determine the actual physical reason for a change in the weather, then attempting to estimate how this will effect the weather at a future date. At first glance this approach would appear to be nearly impossible because of the thousands and thousands of atmospheric, oceanographic, man-made, natural, and extraterrestrial variables which can effect the weather. However, certain variables are more important than others and their mathematical interrelationships have been used to construct viable forecast tools.

The dynamic approach to meteorology, as it stands now, mathematically compares certain key weather variables at a given point in time, then projects this relationship into the future. This technique has proved most beneficial to the meteorologist by providing him with excellent forecasts for the configuration of the upper-level winds or jet stream as much as five to ten days in advance. It is most accurate in the two- to three-day range and, in fact, has become an indispensible tool for the experienced meteorologist. The accuracy decreases beyond three days but still provides a good estimate out to ten days. However, the dynamic approach is not currently capable of providing consistently accurate results for any period much greater than ten days.

There is no question that the availability of accurate three-day projections of upper-level wind or jet stream patterns has markedly increased forecast accuracy. If accurate upper-level wind patterns could be provided for longer periods of time, our long-range forecasting ability would be absolutely astounding. But this isn't the case. That's why we're forced to rely on the statistical methods described earlier if we're to issue long-range forecasts at all.

Hopefully, in the future the dynamic approach will yield the same kind of results for longer periods of time that it now gives in the three-day range. This capability may be long in coming. We've got a great deal more to learn about our atmosphere and how it works. We must also wait for further advances in computer technology which will provide us with the computing ability to handle the massive dynamic equations which will be necessary to provide solutions for days and weeks in advance.

Super-fast computers with fantastically huge data storage capabilities are sure to arrive in the very near future. However, our knowledge of the atmosphere is progressing at an agonizingly slow speed which is not likely to increase in the foreseeable future. What we know about the atmosphere and weather already fills volumes; what we don't know could fill ten times as many volumes. The conclusion is that there's not much hope in the immediate future for accurate long-range forecasts resulting from the dynamic approach.

Hence, at this time, the statistical approach appears to be the only consistently accurate method for producing long-range forecasts. We already know it works and that it works very well for temperatures and reasonably well for precipitation. Table 2 gives you some indication of the accuracy achieved by the statistical outlook method as used by two private sources and the National Weather Service.

Table 2 indicates that long-range temperature forecasts can be phenomenally accurate and, even on the average, are much better than an educated guess. Precipitation forecasts are less accurate and more likely to bust, as explained earlier, and yet are still better than chance.

Another way to assess the verification data in Table 2 is to analyze the number of times each source exceeded or fell below a certain level of accuracy for temperature or precipitation forecasts. This is done in Table 3. It shows that source 1 was by far the most consistently accurate for temperature forecasts and that sources 1 and 2 were nearly the same for precipitation.

EFFECTIVE RANGE FOR ACCURACY

As with short-range forecasts, there is also an effective range for long-range forecasts beyond which accuracy will decrease with time. The greatest accuracy for long-range forecasts, on a monthly basis, will generally fall in the period six months or less from the time the forecast is issued. Beyond six months, unaccountable factors begin to play an

TABLE 2. LONG-RANGE FORECAST VERIFICATIONS

DATE	SOURCE 1		SOURCE 2		NWS	
	Hits		Hits		Hits	
	Temp	Pcpn	Temp	Pcpn	Temp	Pcpn
MAY 1977	71%	50%	73%	36%	63%	42%
JUNE 1977	71%	63%	68%	68%	71%	42%
JULY 1977	92%	62%	45%	64%	100%	42%
AUG 1977	91%	70%	96%	72%	92%	62%
SEP 1977	49%	71%	70%	74%	66%	66%
OCT 1977	78%	47%	69%	65%	53%	53%
NOV 1977	23%	64%	66%	68%	72%	56%
DEC 1977	73%	78%	69%	65%	47%	33%
JAN 1978	94%	52%	17%	80%	22%	69%
FEB 1978	53%	64%	10%	31%	61%	70%
MAR 1978	75%	69%	69%	65%	83%	36%
APR 1978	77%	64%	90%	69%	58%	42%
MAY 1978	50%	58%	55%	62%	75%	50%
JUN 1978	75%	58%	79%	66%	83%	56%
JUL 1978	81%	69%	52%	66%	81%	33%
AUG 1978	75%	61%	66%	62%	78%	58%
	71%	63%	62%	63%	69%	51%

This table gives the verification percentage for two private sources and the national weather service (NWS) for the southern one-third of the nation for temperature and precipitation. The percentage refers to the number of correct forecasts or hits for 30 separate locations.

TABLE 3. ANALYSIS OF TABLE 2

TEMPERATURE	SOURCE 1	SOURCE 2	NWS
Number of Times Verification Exceeded 70%	12	5	8
Number of Times Verification Exceeded 80%	4	2	5
Number of Times Verification Exceeded 90%	3	2	2
Number of Times Verification was Less Than 50%	2	3	2
PRECIPITATION			
Number of Times Verification Exceeded 60%	11	14	4
Number of Times Verification Exceeded 70%	3	3	1
Number of Times Verification Exceeded 80%	0	0	0
Number of Times Verification was Less Than 50%	1	2	7

This table provides an analysis of the various levels of accuracy of the sources in table 2.

increasingly important role and will serve to distort the forecast to an ever-increasing degree as the length of the forecast increases. Hence, a forecast for a single month more than six months in advance rapidly becomes a losing proposition. Yet a forecast for a season as much as several years in advance will often be at least reasonably correct because the factors which rapidly distort a monthly forecast will rarely completely distort a period as long as a season (a three-month period such as fall, winter, etc.).

Currently most, if not all, long-range forecasts for periods more than three weeks in advance are based on the statistical method described in this chapter. These forecasts are essentially a projection of proven cyclical repetitions of weather patterns observed over long periods of time. They are credible and, as shown in Table 2, correct more often than not. Long-range forecasts in the five-day to three-week range are equally as accurate. Hence, this can and should be used as a definite management tool in all businesses dependent on weather.

For example, farmers should be using long-range seasonal forecasts to plan irrigation distribution and purchasing, subsoiling, crop planting, equipment purchases, tree clearing and/or land development, land purchases or sales, and in the development of multi-year business plans. The farmer should also be using long-range forecasts in the five-day to three-week range to time his planting, herbicide and pesticide applications, irrigation scheduling, fieldwork, machinery repair, harvesting, etc.

Business and government now pay hundreds of millions of dollars annually in an effort to produce high quality long-range forecasts. The accuracy isn't bad now and it's going to get better. As in the worldwide technological competition, it is just as obvious to government and business leaders that a sudden spectacular breakthrough in long-range forecasting ability would be a truly significant international event with far-reaching implications. It could mean literally billions of dollars to an individual or company and prove to be a powerful political and military weapon to a government. The stakes are fantastically high, and its entirely possible that such a breakthrough will occur.

Consider the importance of a government's having truly accurate advance notice of the time, severity, and location of harsh winter weather as much as six months to a year in advance. This knowledge could have considerable political and military significance. Severe winter weather in the modern world interrupts transportation, damages winter crops, creates a much greater demand for energy, and affects military preparedness, strategy, and deployment.

Advance notice of significant changes in the global *climate* could prove to be of even greater significance. For example, if the global climate is, in fact, heading towards another ice age, the resultant weather would have a drastic effect on political and military security worldwide. If pronounced changes toward colder weather occurred within the next decade it could result in a drastic reduction of the growing season in the major grain-producing areas of the world. Total crop failures could occur in Russia and Canada where the major grain areas are north of 50° north latitude. Prolonged cold weather, as in the "Little Ice Age" of several centuries ago, could increase the size of the polar ice cap, send glaciers southward, and push the average rain tracks further southward leaving the major grain-producing areas high and dry.

Now that we see the consequences of colder and drier weather in the Northern Hemisphere, we should combine forces on an international scale in a search for better long-range forecasting capabilities. We should check out all sources claiming superior accuracy and determine, once and for all, if there's any truth to their claims. We should harness the power of the most modern computers and search for all possible statistical associations between natural, atmospheric, extraterrestrial, oceanographic, and man-made variables with the actual weather patterns. The statistical approach has already proven successful; further enhancement of this technique should be pursued.

Suppose the weather was magnificent for the next several centuries, resulting in bountiful harvests, unlimited population growth worldwide, and eventually, disaster from an overpopulated world with dangerously few natural resources. Suppose, on the other hand, that the world experienced markedly colder global temperatures within the next few decades or century. This could result in frequent disastrous crop failures, prolonged drought, mass starvation, and heightened military and political tensions, possibly even war. Since we have the capability of supplying respectable ideas about future weather now, aren't the possible consequences reason enough to make a greater effort at supplying more accurate long-range weather forecasts and *using* them?

Part two of this book was written to give the farmer (and others in weather-related businesses) the ability to use weather.

PART II

WORKING WITH WEATHER

5
WEATHER MANAGEMENT ON YOUR FARM

Up to now I've talked only about the meteorologist and what he can do for you. It's time to tell you what you've got to do in order to make the weather information you receive from the professional apply to your farm.

I meet very few farmers who do more about the weather than merely listen to the local weather forecast on radio or television and talk about how bad the forecast is! Farmers have been doing this for generations and most of them don't see any reason to change. "If it was good enough for Grandpa, it's good enough for me."

One of the greatest motivations for change is when your pocketbook begins to look too slim even when you're practicing the same successful farming techniques Grandpa used. Today, profit margins are getting smaller as costs increase. If you'll list all weather-related decisions you make each crop year, you'll soon see that you are spending most of your annual operating capital on decisions significantly affected by weather. Start adding up the dollars you lose every year in lost time, ruined crops, weather damage to machinery, land, buildings, and crop yields. Analyzing the effect of weather on your farm is one of the best places to begin attempting to correct shrinking profit margins.

Very few farmers believe me when I say that you can do something about the weather and actually use it to your advantage on your farm. However, if you'll employ the techniques described in this chapter you'll start seeing results within a couple of weeks and you'll begin to realize that you can actually manage the effects of weather on your farm.

As with any business, if you detect a problem you must first analyze the situation, then correct it, and, finally, use this knowledge to plan the future course of your business more effectively. Is there a problem? If weather affects your yields, you have a problem. Can you correct this problem? Only if you can quantify it, and the only way you can quantify it is to install a weather network on your farm.

A FARM WEATHER NETWORK

The size of your farm weather network will depend on the size and topographical variability of your land. So the very first thing you need to do is to construct a detailed topographical map of your farm, drawn to scale and indicating all major topographical features within the boundary of your farm and for a 25- to 50-mile radius. Indicate the location of all roads, trees, streams, rivers, lakes, and developed or populated areas on or near your farm. Show soil types and irrigated areas plus the crops most likely to be planted in a given area. If possible, go to the extra expense of preparing a bas-relief map of your farm and the surrounding area; this would be an extremely effective tool in selecting the distribution of your farm weather network. It would also be most helpful if you could arrange to have a composite aerial photograph of your farm prepared as a supplement to your scaled map.

An accurate analysis of the physical distribution of your land and the surrounding area is equivalent to a thorough market analysis of the local community by a small business in a city. However, once the market analysis is completed, it may still be beyond the capability of the business manager to construct an effective marketing plan. Therefore, he may still have to consult a marketing specialist or advertising firm capable of putting the market analysis to proper use for his business. The same could also be true for your farm, especially if you're farming very large acreages with considerable topographical variability. You should consider using a consulting meteorologist, well versed in agriculture, to assist you in constructing an effective farm weather network. He should be able to save you the cost of installing unnecessary instruments by optimizing the distribution of instruments to fit the specific requirements of your land, as determined by your scaled map and topographical survey. His consulting fee should be more than compensated by the efficiency of your weather network and your satisfaction with the results you'll obtain.

The first step to installing your network is to determine the location of your base station(s). A base station should consist of at least one set of relatively high-quality weather instruments to measure precipitation, temperature, wind, and any other variable of particular significance to that area of your farm. Each base station should be located in or near each major growing area of your farm and should be conveniently accessible in all types of weather. These are the weather instruments that you should never fail to read 365 days a year. The base station(s) will provide the quality climatological data necessary to make a proper analysis of at least the general effect of weather on your farm. All additional instruments located around your farm will enhance and amplify this base station.

There are many instruments you can choose from, as depicted in appendix B; good choices for collecting precipitation, temperature, and wind, respectively, would be the Official Rain and Snow Gauge, Etched-Stem Maximum/Minimum Thermometer with shield, and homemade Wind Vane with the Hand Windmeter. At the very least, install one quality set of weather instruments near your home or office, whichever is nearer to your major growing areas. You won't be getting all of the information you need, but any information is better than none and one properly constructed base station (see appendix B for further details) can provide a wealth of data.

Installing a base station and a complementary network of weather instruments (as discussed below) are the beginning procedures for initiating weather management on your farm. The term "weather management," as applied to agriculture, refers to the complete process of monitoring and recording the effect of weather on your farm and using this information to save operating costs and market your crops—i.e., to make more money. The philosophy is that you can't manage something you don't fully understand. The most important way to begin is to start collecting the weather data and keep accurate records of crop maturity and all operational functions. You will begin to manage the effect of weather when you can show a direct association between weather, crop maturity, crop yields, and all operational functions dependent on weather.

MEASURING RAIN

The next step in understanding how weather affects your farm is to construct a supplementary set of weather instruments which will allow you to pinpoint specific problem areas. The simplest approach is to distribute inexpensive rain gauges (see appendix B) around your farm. Rain is the single most important weather variable in farming and you should make every effort to distribute the densest rain-gauge network possible. The gauges should be placed in or near each major tract of cultivated acreage. Ideally, each of these areas should be covered by at least three gauges in triangular configuration or a combination of triangular or square patterns. Figure 13 shows a triangular configuration

Figure 13. Sample Rain-Gauge Network

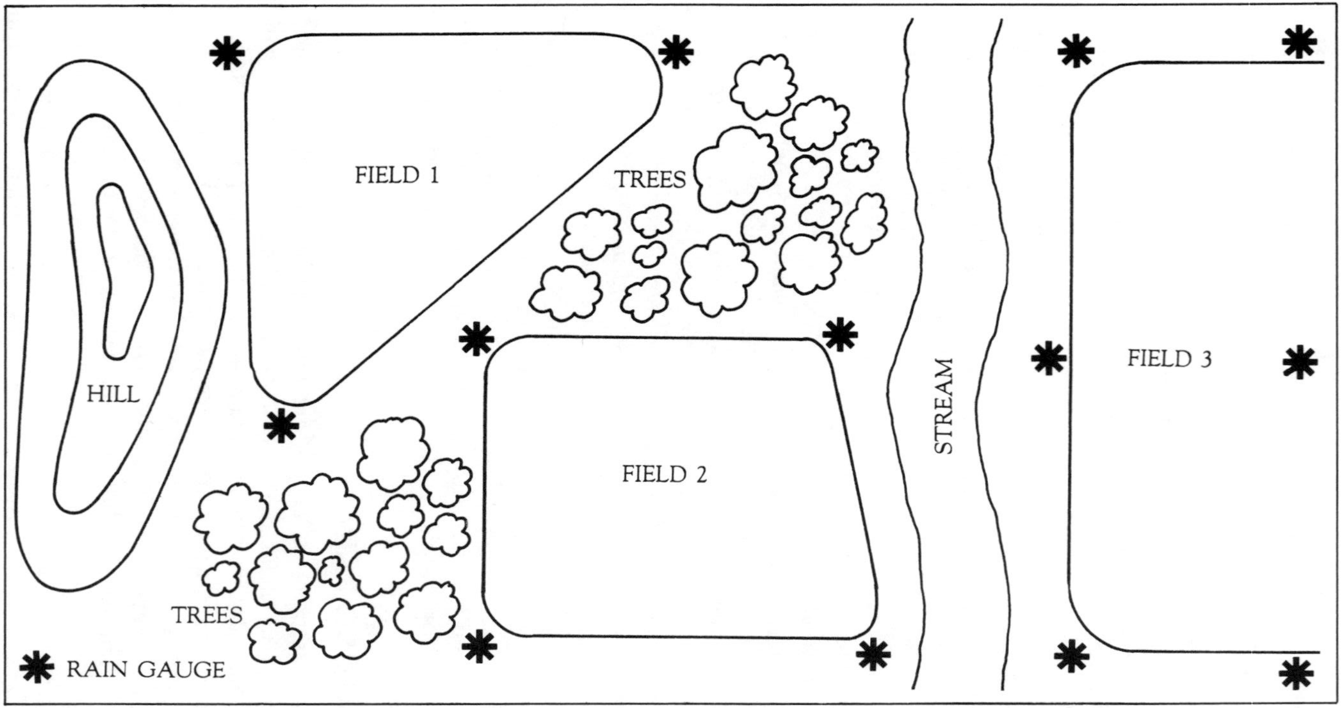

This diagram depicts a sample distribution of rain gauges (designated by an *) around three major areas of planted acreage or fields on a farm.

in field 1, square in field 2, and combination in field 3. Keep the gauges equally spaced from each other and at least 15 feet from low shrubbery or crops and 25 feet from trees. The spacing should be no more than ¼ mile apart. The top of the gauge should ideally be about 3 feet above the ground, but it won't hurt to place them at the top of a fencepost. Place the mouth of the gauge about 3 to 4 inches above the top of the post.

Rain gauges come in many shapes and sizes (see appendix B). Perhaps the best gauge for general field distribution is the homemade gauge shown in appendix B. It is inexpensive, easy to make, very durable, and quite easy to replace. It will also withstand freezing weather reasonably well. You should try to measure all types of precipitation, including frozen precipitation during the winter; this will be an important part of a complete set of weather records for your farm. However, because precipitation is often more general during the winter, making one measurement at each base station would suffice; besides, when the weather really gets bad it will be hard enough just to accomplish that. Snow is measured by pushing a ruler or yardstick through the snow to the ground or surface and reading the depth; it is best to take at least three readings in different locations (away from drifts) and average the depths. The depth should be converted to equivalent inches of rain by dividing by 10; the general rule of thumb is that 10 inches of snow are equal to one inch of rain.

Installing your rain gauges is relatively easy. Now comes the hard part: you must discipline yourself (or someone else if you can afford it) to read the gauges at the same time after each rain. To be truly useful, you should read your gauges at the same time the official National Weather Service gauges are read—at precisely 1200 and 2400 hours, Greenwich Mean Time. For the Eastern Time Zone, that would be 7 a.m. in the winter and 8 a.m. during the months of Daylight Saving Time. In the Central Time Zone, winter readings would be at 6 a.m. and summer readings at 7 a.m.

Be sure to encode all weather observations, including rain, faithfully and neatly. You can use the forms shown in appendix B as a guide. This information is so valuable that you should make every effort to preserve it properly.

ISOHYETAL MAPS

If you construct a dense rain-gauge network, you will need to construct isohyetal maps after each rainfall or storm. Isohyets are lines of equal rainfall; by drawing these lines on a map as shown in figure 14 (solid lines), you will be able to see easily the pattern of precipitation on your farm. The technique for drawing isohyets is called interpolation, a

FIGURE 14. SAMPLE ISOHYETAL/ISOTHERMAL CHART

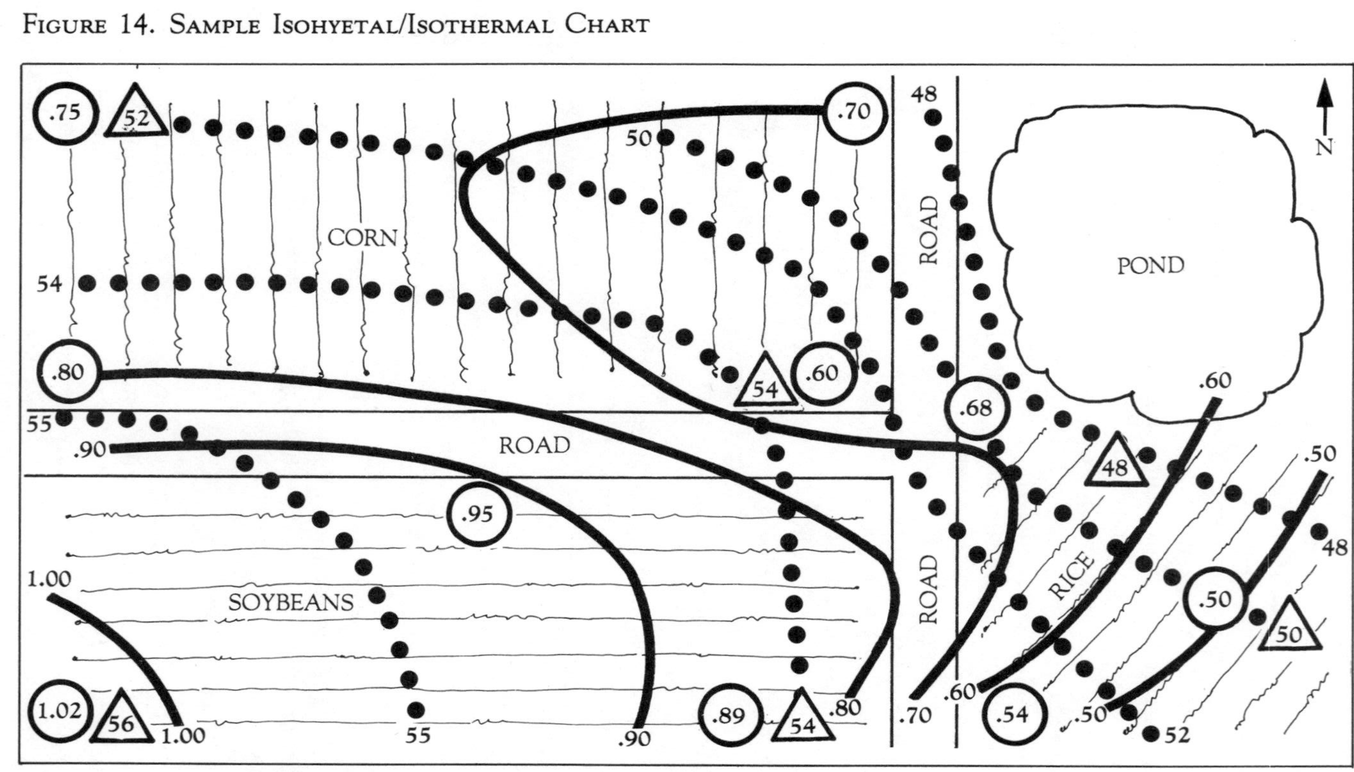

This diagram depicts isohyets (solid lines) and isotherms (dotted lines) for the values indicated as determined from the 24-hour rain totals (circled) and minimum temperatures (triangles).

scientific term for splitting the differences or approximating the value of a given variable, such as rainfall or temperature, between two points at which that variable is actually measured.

In figure 14, the circle indicates the location of a rain gauge and the triangle, the location of a maximum/minimum thermometer with appropriate shield (see master map, figure 61). Assuming that the weather instruments were checked just after dawn, the sample readings for rain (in inches) and minimum temperature (°F) are shown in the appropriate space. For example, there was .75 inch of rain, and the minimum temperature was 52°F in the northwest corner of the corn field; .50 inch of rain and a minimum temperature of 50°F was observed on the eastern edge of the rice field.

In order to draw the isohyets, first analyze all the rain data and you'll see that the rainfall collected ranged from a low of .50 inch in the rice field to 1.02 inches in the soybean field. Drawing isohyets for each tenth of an inch from .50 to 1.00 and starting with .70 inch, you can see that the .70 isohyet can be neatly sketched starting from the northeast corner of the corn field, splitting the difference between .80 and .60 on the south side of the corn field, running about one-third of the way between .60 and .95, going almost up to .68 by the pond, then smoothly curving southward and splitting the difference between .89 and .54. (This would mean that you would expect .70 inch of rain along the entire length of this particular curving line.) A similar procedure would be followed in drawing the remaining isohyets and isotherms (dotted lines).

Isohyets and isotherms, as shown in figure 14, will not indicate the exact precipitation or temperature for a given location but they'll be a very good approximation. The denser the network, the better the results and the more complete your analysis will be. In figure 14 you can be fairly confident that the precipitation at any point along the .70 isohyet was actually equal to or very close to .70 inch. If the pattern of isohyets in this diagram persists for the majority of storms, it would mean that the farmer had wisely planted his corn and soybeans in areas likely to receive the most rainfall, while his irrigated rice acreage was planted in a drier area.

You already know what effect weather generally has on your farm, but the isohyetal maps will provide you with the kind of detailed data you really need. You should almost immediately begin detecting marked differences in rainfall from one field to the next. You should also detect that rainfall on your farm is usually in a definable pattern, with certain fields consistently receiving more rainfall than others. The pattern of rainfall will change gradually from year to year and season to season, but consistent patterns should still be detectable every year as long as there have been no drastic and recent changes in the local topography or vegetation, such as rows of shrubbery or trees or forested areas. The result will be a pattern of consistently wetter than normal, or "wet," fields and drier than normal, or "dry," fields on your farm.

One use for this information can be illustrated by analyzing the effect of a forecast for a 50% chance of showers in your area. You already know after reading chapter two that a forecast for a 50% chance of showers means that it is definitely going to rain but only over 50% of the area. You equate this to your farm by saying that if the forecast holds true you have an excellent chance of receiving rain on your "wet" fields, such as the soybean field in figure 14, but the odds are not favorable for rain on your "dry" fields. This could mean big savings for the farmer who, after hearing the forecast for a 50% chance of showers, decides not to apply expensive pesticide on a "wet" field.

Where your fields are in respect to rain tracks or patterns can make a big difference in crop yields. The location of rain tracks should be a factor in your decision of which crops to plant and where you'll plant them. Don't rush into this decision; collect at least one full year's worth of data before making major changes in your planting practices. You will find that the tracks will vary at least a little in each season of the year because of changes in the prevailing upper-level flow or storm track. Occasionally, particularly in the South, storms will move from east to west under a strong flow of upper-level tropical winds; rain tracks would not apply in this situation.

In summer (again in the South), upper-level tropical winds will occasionally migrate northward over the Gulf of Mexico; easterly waves, or pockets, of cold air imbedded in this flow can produce hurricanes or large, loosely organized storms. The latter will sometimes drift westward across the Southeast and into the Mississippi Valley. Because storms rarely come from this direction, the resulting rains will not fall in the same patterns normally seen with the more common movement of storms from west to east.

Another example of when rain will not adhere to its normal patterns occurs in the fall and spring when large upper-level low-pressure systems will become detached from the main jet-stream flow. These detached or "cutoff" lows are a meteorologist's nightmare as they leisurely drift eastward and sometimes loop up and around the entire length of the Mississippi Valley before slowly migrating eastward again through the Southeast.

Another possible use for the information you'll glean from your rain-gauge network would be in selling or leasing your property. If you could show proof of consistently beneficial rainfall on a particular field, you could sell it or lease it for more money. Just don't let the county tax assessor in on your little secret or you'll be paying higher taxes!

You'll get even better information if you can convince your neighbors—whether they farm or not—to put out gauges. Weather data for agricultural areas across the nation is so sparse that any additional data is useful. Your gauge(s) will be of enormous importance to you and to a meteorologist analyzing the weather for your farm and your county. However, if several people in every agricultural community began keeping accurate rainfall records, it would make the record just that much more complete and valuable to every member of the community. When agriculture is an important source of revenue for a community, rain gauges and other weather instruments used to monitor the weather should be a mandatory project for every farmer and every member of the community whose livelihood depends on agriculture. Living in ignorance of the specific effects of weather may be a blissful experience, but, for an agricultural community, it's not very profitable!

TOPOGRAPHICAL CHANGES CAN CHANGE THE WEATHER

Your rain-gauge network will give you a better idea of how rain is affecting your farm, but it may also be useful in other ways. For example, if you have a tract of land which at present is receiving enough rainfall for good crop production, rainfall information might convince you or a neighbor not to clear a nearby large acreage of trees. Trees are a source of moisture to the atmosphere and they alter wind currents. Removing them could change rainfall patterns.

For many decades now the philosophy in most agricultural areas has been: when in doubt cut the trees down to clear more acreage and make way for ever-larger machinery. The U.S. Corps of Engineers has also contributed to this philosophy by clearing massive acreages of trees along streams and rivers. Yet, if you'll ask almost any farmer who has lived in one area for a long time, he'll be able to recall when rainfall patterns have seemed to change suddenly and for no apparent reason. He'll undoubtedly also be able to recall that the removal of a large acreage of trees or similar drastic change to the local topography was coincident with the change in rainfall. Rainfall patterns will change gradually as the climate varies but a sudden, drastic change will almost always imply a significant change in tree cover or similar topographical alteration.

You should always be careful in removing large acreages of trees or prominent rows of hedges or shrubs. Each of them contributes to the general pattern of weather for your local area. Removing any one of them could be a mistake.

Of course, similar changes in the local climate can be caused by other major drastic changes in the local topography: large construction projects, new roads, new dikes, large earth-moving operations, etc.—any major change in the surrounding land configuration could induce a change in local weather patterns.

By closely monitoring at least the normal distribution of rainfall around your land, you'll be much better informed about the local climate and less likely to make, or allow someone else to make, a large-scale removal of trees and shrubs or a drastic change in the local topography. Not only will you be better informed but you'll be giving your meteorologist more information to work with.

In America we have the most sophisticated weather-gathering network in the world. But it isn't enough to give meteorologists the detailed information needed to make complete assessments of how weather is related to conditions in small areas like single counties—much less individual farms. Rain gauges in your fields, and hopefully your neighbors' fields as well, could answer some critical farming questions, save you a significant amount of money in a short time, and help fill in the gaps for your meteorologist.

Your network of rain gauges will begin supplying important information immediately and the value of this data will increase with the years. The longer you collect weather data on your farm, the more complete the picture of the effect of weather on your farm will become. Several years of data will reveal subtle shifts in the pattern and even cyclical patterns of wet/dry and cold/hot variations.

Every farmer should consider the installation of a dense network of rain gauges on his farm. The farmer who irrigates should do even more. You not only need to know how much rain has fallen and where, but also how quickly it is evaporating from the soil or draining off underground. If you attempt to schedule your irrigation application merely by observing plant stress or how dry the soil looks, it will usually be too late or too early to irrigate. You must quantify soil moisture just as you do precipitation.

MEASURING SOIL MOISTURE

Evapotranspiration, or loss of moisture from the soil around your crop, can be easily measured by installing at least one evaporation pan in or near the crop area. A sample pan, with appropriate measuring device, is shown in appendix B. The pan should be mounted on a convenient platform near the surface and in or by the irrigated acreage.

Soil moisture can be determined in several ways. The most accurate method of all is to take actual core samples, from selected locations within the irrigated acreage, to the depth of the roots. Each core is weighed, then baked in an oven to evaporate the moisture, and weighed once again; the weights are compared to determine the amount of moisture lost and, hence, the effective percentage of moisture to the depth of the sample. This process is perhaps the most flexible and inexpensive but it is far from convenient.

A more convenient method for determining soil moisture involves the measurement of resistance in electrodes, or soil blocks, buried to various depths in the soil up to the root depth of the plant. The buried soil block releases or absorbs moisture until its contents approach equilibrium with the moisture content of the soil. When the exposed lead of the soil block is connected to a meter, the current flow between the electrodes will indicate the amount of moisture at that depth. Soil blocks can be buried at various depths and locations across the irrigated area and conveniently read with a portable meter. This is by far the most convenient and economical method in the long run for direct on-site measurement of soil moisture in irrigated areas. The instruments used in this method are shown in appendix B.

One final method for determining soil moisture is by mathematically correlating actual rainfall with actual soil moisture over a period of time. This information can be used to construct a graph or table which will give you a good approximation of the amount of time it usually takes to reduce the soil moisture to the point at which irrigation is necessary. The amount of rainfall, as measured in your Farm Weather Network, can be compared to the mathematical data to determine when you should schedule your next irrigation.

The best method of measuring soil moisture on your farm will probably be a combination of several techniques. You should experiment with each of the above methods and select the ones that work best for you. Keep accurate records by encoding your daily observations in your daily weather log as shown in appendix B. It is a fairly good assumption in most cases that you've waited too long to irrigate by the time you observe plant stress.

You will also need to determine just how much moisture you're going to maintain, and to what depth at each stage of maturity, for the crop in question. You should consult a local agronomist or plant specialist to help you determine the amount of soil moisture that's right for your crop and your soil.

Since rain and soil moisture are not the only factors affecting crop yields, you should also consider at least two other weather variables potentially important to crop yields: temperature and wind.

MEASURING TEMPERATURE

You should place thermometers that measure maximum and minimum temperatures on each tract of land. Be sure to attempt to record the daily maximum and minimum temperatures in or near each major variation of the topography within your planted acreage. The distribution of your thermometers will not be as uniform as your rain-gauge distribution, for you are monitoring only significant variations within the crop area. Maximum/minimum thermometers are not hard to find and they are relatively inexpensive—no more than $20 to $30 for the inexpensive models (see appendix B). However, depending on the variability of your land, you shouldn't need more than two or three of the less expensive thermometers per major tract of planted acreage.

What you're looking for are significant variations in temperature caused by topography, soil types, prevailing wind, crop condition, and other factors of a semi-permanent nature. Are these sustained or frequent variances in temperature affecting crop maturation and yield?

In order to assess the effect of temperature on your crops accurately, you will need to install a relatively dense network of thermometers around your major crop areas. You can construct isotherms from these observations as shown in figure 14. The pattern of temperature in that diagram shows a definite pattern of cooler temperatures near the pond. This would imply that there is a definite slope to the land, with the lowest area being around the pond. Cold air, being heavier, will sink to the lowest areas at night; this phenomenon is called cold air drainage (see figure 15).

Is the nightly drainage of cold air onto your planted acreage sufficient to reduce yields in at least a portion of your acreage? Research has shown that cold air drainage occurs everywhere and on nearly every night of the year except during a change in the weather. This is one reason why fog frequently forms in creek and river basins at night; cold air drains into the basin, quickly drops the air temperature to the dewpoint, and condenses the available moisture in the air, resulting in fog. It would be logical to assume that dew formation would be more consistent and heavier on crops situated in low-lying areas affected by cold air drainage.

What effect does cold air drainage have on crop yields? Can you quantify this effect? Are your crop yields affected enough to warrant monitoring of this phenomenon? You can't answer these questions until you actually record daily temperatures in your crop areas and compare yields. If the differences are significant enough, you might be able to take advantage of the situation by altering crop selection, planting dates, seed type, etc.

FIGURE 15. SAMPLE MAXIMUM/MINIMUM THERMOMETER DISTRIBUTION

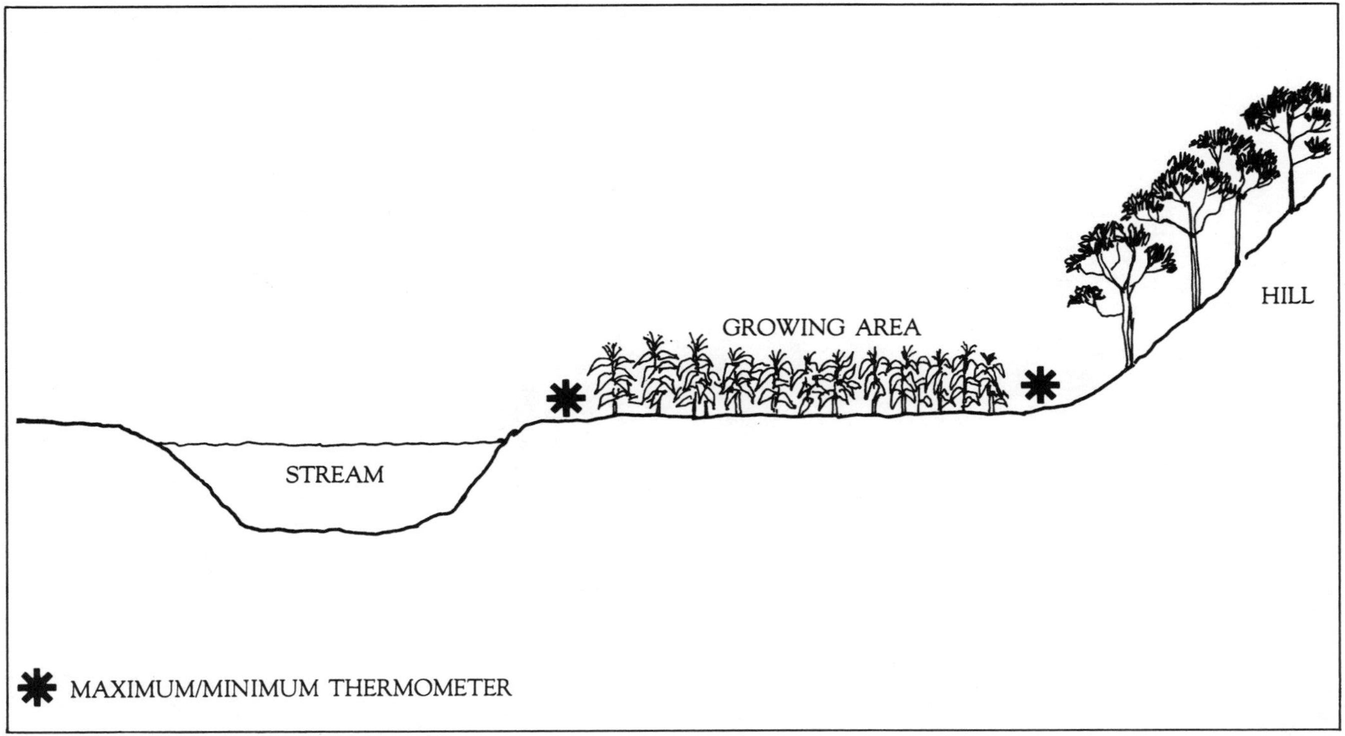

STREAM
GROWING AREA
HILL
✱ MAXIMUM/MINIMUM THERMOMETER

This cross-sectional diagram depicts a sample distribution of maximum/minimum thermometers strategically situated so as to monitor the effect of cold air drainage near a growing area.

MEASURING WIND

As with temperature and precipitation, few people are aware of the specific effect of wind on crops. We've all heard of farmers who've lost entire crops when a strong wind essentially sandblasted the young seedling, shredding it to bits.

A good example of this problem occurred in 1978 on a farm in the Mississippi Delta. A farmer had just planted over one-thousand acres of cotton. A large rain followed soon thereafter and soaked the ground. Before the farmer could get back in the field to break up the ground, a strong wind lifted fine dust and sand particles off the moist soil and completely sandblasted the young cotton seedlings to shreds. The crop was a total loss.

How many farmers do you know who practice any sort of wind management? Can anything be done about it anyway? Some Chinese farmers erect windbreaks alongside growing areas to protect the young plant just after it's broken through the surface. They use several parallel rows of woven reeds slanted in the direction of the prevailing wind to protect the young seedling. As the prevailing wind hits the windbreak it is chopped up and slowed considerably as it crosses over the young crop; sand and small grains of dirt, potentially harmful to the crop, either fall harmlessly to the ground or are lifted up and over the crop.

In order to construct a windbreak, you will need to make your airfoil at least three feet high. You can use reeds or any other material that can be latched together to form a portable barrier to the wind. Slant the windbreak at about a 45° angle directly into the wind. One barrier will be effective, but two or three windbreaks in a parallel formation should prove even more effective.

Here again, it will be difficult for you to know how to apply aerodynamics or wind management to your farm if you're not aware of how wind is specifically affecting your farm. You should monitor the wind for each major tract of planted acreage just as you are monitoring temperature and precipitation. The best way to do this will be to construct simple weather vanes and collocate them with either your rain gauges or thermometers, or both. You can measure the speed of the wind as you check your thermometers. Record both the speed and direction (see appendix B).

Currently, there are no published standards on wind management. So, you'll just have to experiment. If your crops have been damaged by wind in the past, you might want to construct a windbreak similar to the Chinese kind or you might find it more convenient to plant a row of fast-growing trees on the windward side of the fields most affected. You might even have success with a strip of winter wheat on the windward side of the field. Whatever you do, your goal is to chop up the windflow enough to protect the young seedling.

However, the young seedling stage isn't the only time your crop can be damaged by wind. Winds are most likely to be at their strongest in spring and fall. In the fall, your crop is taller and carries your entire profit for the year. It's not going to do you much good if stalks are broken by the wind and your profit's lying on the ground in several inches of mud. Rice is particularly susceptible to winds in the critical last couple of weeks just before harvest. The top-heavy plant is easily blown over. This can create extreme problems for harvest, particularly when the strong wind is coincident with rain, as it often is!

You may find that the same techniques you used for the young seedling won't apply for the taller, mature plant just before harvesting. Hence, it might be more economical and a more permanent solution to wind management to consider narrow bands of trees or shrubs. Since there are no guidelines you will have to experiment, but the results should mean more money in your pocket and one less worry about one more so-called uncontrollable element of weather.

Don't overlook the fact that as you attempt to employ wind management on your farm, you will be making at least minor changes in the general pattern of temperature and precipitation as discussed earlier. So it is important that you carefully and closely monitor these variables in those areas where you employ various types of windbreaks. Any alteration of the local topography is going to make at least some change in the local climate. A single row of trees may not seem all that important but it could make a bigger difference than you think. So whether you're clearing or planting trees, using windbreaks, or otherwise altering the local topography, your Farm Weather Network will help you keep track of the resultant variations in the weather patterns on your farm. The greater the alteration, the denser you should make your network in order to construct the most accurate record possible of the resulting changes in the weather patterns commonly observed.

By employing the techniques just discussed, you will be taking the initial step into weather management. You may learn more about your farm than you ever suspected. By knowing your farm better, you will be in a position to start actually controlling the effect of weather on your farm.

KEEPING RECORDS

As mentioned earlier in this chapter, you should make every effort to keep a neat, orderly record of all weather data. This can't be stressed enough, for the only real value of having a weather network comes from accurate record keeping and the detailed analysis of these records to detect the most consistent or significant effects of weather on your farm. It would also be a great idea to make a good carbon copy of every record. A second copy of your valuable weather records will insure their safety just that much more and could also save countless hours of transcribing if your records are forwarded to a consulting meteorologist for further analysis. Never discard any records, for weather records, like fine art, become more valuable with age. See appendix B for a more detailed discussion and a sample log form to use for encoding your daily observations.

FIGURE 16

CROP MATURITY LOG: Cotton	
DATE	SIGNIFICANT STAGES OF MATURITY
April 27-May 4	Plant
May 9	80% of crop at a stand
May 12	100% of crop at a stand
June 12	90% of crop at first square
July 20	90% of crop at first bloom
Oct. 1	70% of crop with open boll
Oct. 15	100% of crop with open boll

COMMENTS: CROP YIELD DATA:

Field One - 520 pounds per acre
Field Two - 490 pounds per acre
Field Three - 580 pounds per acre

Once you've begun to collect daily weather data, you must then begin to keep an equally complete supplemental record of all important stages in the life of your crop(s) and all operational functions you must accomplish in the maintenance of your crop(s), livestock, and farm. These are the records that will enable you or a consulting meteorologist to show conclusively the relationship between weather and all of the products you produce and functions you must accomplish in order to show a profit at the end of the year.

The first supplemental record you should keep is entitled the Crop Maturity Log and is shown in figure 16 with sample entries included. This is the log in which you list all stages of maturity for each crop you produce. List everything from planting dates to first stand, each significant stage of development, harvest dates, and yield. Whenever possible be sure and specify the differences in yield from field to field to enable you better to define the effect of weather on your crops in specific areas. You should also list any special problems you encountered during the growing season which might possibly prove to be of significance when searching for associations.

One of the reasons crop yield data is so important is that it can be used to construct crop yield models. Using long-term records of weather and crop yield data, you can show an association between temperature/precipitation and crop yields. This statistical association can help you make a very accurate estimate of the potential yield of your crop at any stage in its development. This knowledge, plus a reasonably accurate long-range forecast through the remainder of the growing season, can help you to market your crops better by having better knowledge of their potential.

The second supplemental record you should keep is a log of all operational functions necessary in the maintenance of your farm, crops, and/or livestock. Don't neglect any function in your log, for the majority of your work can be affected by the weather. This record will help you to quantify that effect and give you a better idea of just how much weather is affecting your pocketbook. The Farm Operations Log shown in figure 17 has several sample entries to give you an idea of what you should record. In the "comments" section at the bottom of the page, list any special problems you encountered while performing these functions.

Whenever possible, attempt to show how much money was lost or saved because of the effect of weather on the development of your crop or a particular operational function. For example, how much would it cost if you were forced to replant because of torrential rains and cold temperatures immediately following your initial planting? This problem, of course, haunts all farmers and, in some years when the planting weather

FIGURE 17

FARM OPERATIONS LOG	
DATE	OPERATION PERFORMED
March 15-19	Disc
March 16-20	Chisel
April 10-14	Incorporate herbicide
April 18-22	Disc
April 23-28	Harrow
May 15-June 2	Plant
June 7-14	Cultivate
June 22-30	Cultivate
July 15-22	Cultivate
July 24-28	Apply insecticide
Oct. 15 - Nov 3	Harvest

COMMENTS:
Forecast 50% chance of showers July 27th; last insecticide application delayed to 28th. Potential savings of $3,000. Bad weather forecast for first week of Nov. so used custom pickers last 7 days of harvest. Bad weather came Nov 7, 8 & 9. Potential savings of $10,000.

is particularly nasty, can create losses of several hundred million dollars nationwide. Harvest weather is also extremely critical to every farmer, for your entire operating revenue plus any profit can agonizingly wilt before your eyes when disastrous weather attacks your crop before you've had a chance to harvest it. Other visible losses can occur when rain washes off a herbicide or pesticide before it has a chance to work. By the time you calculate all such losses at the end of the growing season, the figure may well astound you. You will very easily prove to yourself that something must be done to protect yourself as much as possible from big losses due to weather.

USING WEATHER FORECASTS IN PLANNING

Once you have begun to analyze the effect of weather on your farm and associate this with the maturity of your crops and your operational functions, you then need to start using this knowledge to your benefit. This can be conveniently accomplished by constructing a worksheet, as shown in figure 18, showing the five-day forecast and the functions you expect to accomplish on that day. You can get a five-day forecast from several sources, as discussed in chapter two, and tailor it to your farm by using your daily weather log.

Unless you consult directly with a meteorologist, you will have to tailor the forecasts for figure 18 yourself. This can be done by carefully comparing your daily observations to the nearest official observation site from which the forecast you use is disseminated. For example, your maximum and minimum temperatures should show a consistent difference from the official readings because of topographical differences between your farm and the official observation site. You can usually subtract or add this difference, whichever the case may be, in order to tailor the temperature forecast to your farm. This procedure will be accurate most of the time. You can tailor the precipitation forecast by observing how frequently you receive rain on your farm, or on a particular field, and by noting the number of times you got rain versus the number of times it was forecast. For example, if out of the last 10 rain systems, your corn field received 8 rains, you should assume that whenever a reasonable chance for rain is forecast your corn field is likely to get rain.

By getting in the habit of preparing a Five-Day Plan every day, you should begin to realize a definite savings in both time and money. Your daily analysis of weather forecasts from several different sources will help you to determine which source is best for your planning needs. This knowledge, plus the knowledge gleaned from your daily observations, will definitely make you a more efficient farmer.

FIGURE 18

FIVE-DAY PLANNING LOG					
DAY/DATE	MON JUN 1	TUE JUN 2	WED JUN 3	THU JUN 4	FRI JUN 5
WEATHER FORECAST	Sunny	Sunny	Sunny	Cloudy slight chance of showers	Moderate to heavy rains
OPERATIONAL FUNCTION	Cultivate	Cultivate	Cultivate	Cultivate Machinery repair Crop Charting	Machinery repair Record keeping Crop Charting
COMMENTS: Next best chance for rain on Tuesday, June 9th					

Your use of weather forecasts shouldn't stop with a five-day forecast. As discussed earlier in the book, long-range forecasts for periods ranging from several weeks to several months and years are available. The National Weather Service, in the *Weekly Weather and Crop Bulletin,* issues both monthly and seasonal forecasts for temperature and precipitation. Several reputable private weather services also issue monthly and seasonal forecasts in periodicals and by direct mail. To get even longer-term forecasts, you'll have to consult a professional meteorologist who can construct a chart for your area showing the normal cycle of temperature and precipitation over a period of several decades.

You should not ignore these products, for they are quite capable of giving you a very good idea of what to expect in the way of weather, in the future. You should be using long-range forecasts to construct a five-year plan for your business. A knowledge of the most likely weather to occur through the five-year period will help you to plan purchases of irrigation equipment, land, machinery, etc.

For example, if your summers are currently in a hot/dry weather pattern which is likely to persist for the next five years, you might want to consider purchasing irrigation equipment for each of the next five years, then taper off your purchases as summers get cooler and wetter. Or, because some crops favor hot/dry weather more than others, you might consider growing more of one crop than another through the next five years and tailoring your equipment and machinery purchases accordingly.

As you work through the recommended steps in this chapter, it doesn't take long to see that the potential earnings through applying weather management to your farm should be well worth the effort. In some cases the savings can be phenomenal, especially in those locations where there are frequent pesticide applications. You'll lose less of these applications because of what your weather network has revealed to you about your farm and your ability to use this data to tailor forecasts to your farm.

The return on your investment in a Farm Weather Network will be proportional to the effort you expend in constructing and maintaining it. Cost will not be a significant factor, for a simple but effective weather network on most farms can be installed for less than $500, including maximum/minimum thermometers, rain gauges, and wind vanes. Reading and recording the data will be the hardest part, but even that can be relatively inexpensive if it's a family project or you pay some enterprising young boy or girl to do it for you.

It seems odd to say that the introduction of weather management to farming is a radical idea. But it is. For the effect of weather on farming is usually accepted stoically and just considered as one of the hazards of the business. You'll spend hundreds of thousands of dollars on land, equipment, seed, and chemicals, but only five dollars for a cheap rain gauge in your backyard and a thermometer outside the kitchen window. Yet despite your enormous investment in the standard farming necessities, weather is still going to be the deciding factor in crop yield. It's about time you started doing something about it!

You observe and mentally catalogue the effect of weather on your crops but until you scientifically monitor this effect, you'll have little real knowledge to work with. It is very difficult to manage something that you don't fully understand. By distributing a Farm Weather Network as described in this chapter, you'll have manageable data to work with. You'll be able to start using weather management and stop completely unnecessary losses in crop yields due to weather.

For now, most of the improvements in weather data, potentially of benefit to you, are going to have to start on your farm. But, as recommended earlier, you should encourage all of your fellow farmers and neighbors to begin collecting weather data and keeping good records for future reference. Hopefully, every agricultural community in America will someday make observations and keep accurate records as an integral function of that community. Then all farmers will be in a better position to "manage" the effect of weather on their farms.

6
THE APPLICATION OF WEATHER MANAGEMENT TO CROP MARKETING

In my opinion, the two most important factors affecting a farmer's pocketbook are weather and marketing. You've seen how weather management can be applied to your farm, but it is equally important that you also know how to apply weather management to the marketing of your crops. The two, weather and marketing, may, at first glance, not appear to be related. But you know that weather directly affects your crop and farming operation and that the marketing of your crops determines the final price you receive at the marketplace.

You are already well informed about the technical processes of marketing your crops. You know about the futures market, cash market, and the comparison of the futures price to the local cash price to determine the basis. This is a part of your repertoire as a farmer in the modern world. But, what you may not be aware of is the significant role that weather—weather almost anywhere in the world—plays in determining the price of crops at the Chicago and New York commodities markets and, consequently, the price you receive. However, by simple projection, it is quite easy to see that the considerable effect of weather on your crops multiplied millions of times to include all crops, worldwide, is an imposing force indeed in the yields and, consequently, the price of crops worldwide. By understanding the role that weather plays in this process you can begin to employ weather management in the marketing of your crops.

Every farmer in America is a member of the world community of farmers. What farmers around the world produce directly affects the price you receive. Hence, you have to be aware of the dynamic forces affecting crops worldwide in order to market the crops you produce in your small section of that world.

WEATHER AND THE COMMODITIES MARKETS

The prices of crops on the Chicago and New York commodities markets are fundamentally affected by supply and demand. When supply exceeds demand, prices will generally fall. When demand exceeds supply, prices will generally remain firm or go up. There are many factors involved but one of the most fundamentally important factors affecting supply and demand is weather. Normal weather will usually result in bumper crops and a greater supply than demand. Abnormal weather (too hot, dry, cold, or wet) will generally result in decreased yields and greater demand than supply. Hence, over the length of a growing season, weather will always have a direct effect on price because of its direct effect on crop yields.

In 1978, for example, the weather was abnormally dry in the High Plains of Texas and Southeast Arkansas, which resulted in very poor yields for non-irrigated cotton. This contributed to a decrease in supply and prices remained stable or rose slightly higher. Here was a perfect example of the direct effect of weather on prices. At the start of the 1978-79 growing season, the price for cotton was hovering around a meager 50¢ a pound and the prospect for higher prices was bleak. Then came a disastrously cold, wet planting season in the Mid-South followed by drought in the High Plains of Texas and Southeast Arkansas. The end result was a cotton crop fully one-fourth smaller than the previous year. This drastic decrease in supply was the major reason cotton prices surged towards 70¢ a pound by the end of the harvest season.

In the early 1970s, blight struck the corn crop in the Midwest one year, and prices soared because of decreased supply. The following year blight was expected again but a hybrid corn plus very favorable weather resulted in a bumper crop and prices dropped. This was exactly the opposite of what was anticipated, because the weather was expected to be equally as favorable to the spread of blight. However, the hybrid variety so reduced the effect of blight that weather did indeed become the dominant factor which resulted in the high yields.

Weather also plays an important role in affecting crop prices on a much shorter term by hampering both the supply and harvesting of crops. In winter, cold weather over the Upper Mississippi Valley will freeze the Mississippi River; the farther south it freezes, the greater hindrance it becomes to barges laden with soybeans, corn, and other agricultural products. This temporarily reduces supply and increases demand for those products, which results in a higher price; this may create a short-term favorable situation for the farmer in lower portions of the Mississippi Valley who can still deliver his products by barge to other markets, such as New Orleans.

This situation developed in January and February of 1978 when the Mississippi River was frozen as far south as Cairo, Illinois. Farmers in the Mid-South were suddenly faced with an improved price outlook, as the basis strengthened in response to the supply problems in the Main Corn/Bean Belt generated by the frozen river. Several long-range forecasters had foreseen this bitter trend in the weather several months or more in advance. A few farmers, who were aware of this, held their crops in anticipation of the bitter cold weather and were able to get a few extra cents per bushel.

Another example of a short-term effect of weather on the commodities markets occurs in both spring and fall. Poor planting conditions can temporarily drive prices higher because of the poorer prospects for a good crop. For example, in the Mississippi Delta, April of 1978 started off with beautiful warm, sunny weather which led cotton farmers to plant early, just before torrential rain and cold weather set in through the remainder of April and the first part of May. This resulted in cotton prices moving higher, as nearly 80% of the entire crop had to be replanted and prospects looked bleak for a good crop. Ironically, though, the weather reversed itself quickly and was ideal through the growing season, resulting in a fine crop for most of the Delta, with the exception of Southeast Arkansas (as mentioned earlier).

In the fall, weather can also create a short-term upward surge in market prices by unwanted rains or snows that temporarily interrupt the harvest. A large rainstorm moving across the South in October, during the cotton harvest, will usually result in a short-term surge in cotton prices. A similar storm over the Main Corn and Bean Belt in the full swing of the harvest will also result in a short-term increase in corn and bean prices.

It is very easy for you to take advantage of such a situation by keeping informed about any major storms which might threaten the harvest areas. As the weather materializes you can hold your crop; then, if prices are acceptable, you can capitalize on a short-term surge in prices coincident with the interrupted harvest due to bad weather. Your timing must be precise, for the surge will often last only a day or less. This is an easy way to make a little extra profit, by taking advantage of the fact that demand is usually at a peak during harvest; any interruption in supply, however small, will quickly create an increased demand and offer of a higher price on the Exchange for the crop so affected.

Obviously, weather plays a major role in both long-term and short-term fluctuation of crop prices on the Chicago and New York commodities markets and, hence, the amount you receive for your crops. The better informed you are about current and forecasted weather, the more you will be able to take advantage of crop prices which fluctuate directly because of weather. Once again, though, it's not nearly that simple—there are just too many other factors involved to approach it on such a simple basis.

THE COMPLEXITY OF THE COMMODITIES MARKETS

The complexity of the commodities markets and the danger of relying on a single input such as weather as your forecasting tool results from both the players involved and important outside factors. Basically, there are four primary participants in the commodity market: speculator, cash merchant, local professional, and farmer or hedger (see Glossary). The term "speculator" includes both sophisticated multimillion dollar firms with massive computers and rank amateurs playing the markets like a one-arm bandit in Las Vegas. The cash merchant is the one who is usually the end recipient of the raw product and either processes the product as well or sells to others who will process the product accordingly. The local professional is a professional commodity trader, a member of the Exchange, dealing (for his clients' accounts only) in the commodities markets. The farmer was listed last because, even though he's the one who is actually producing the commodities for which the trade exchanges were formed, he is, ironically, usually the least well informed about those very processes which ultimately affect the price he receives.

The difference between the first three participants and the farmer is the level at which they participate. The first three (with the exception of the amateur speculator) spend every hour of every working day totally immersed in the entire scope of factors affecting the market, from weather to politics to any and all internal and external factors affecting the market. They have at their disposal massive computers, sophisticated information-gathering and communications equipment, technical analysts and chartists, and lobbyists who maintain constant contact with important political and administrative figures involved with agriculture. The farmer cannot and should not attempt to participate at this level because he is quite simply outgunned by superior forces. The farmer must not try to be a speculator. At the same time, however, he can still participate effectively at his own level from which he ultimately controls the cards, for the high-class professional participants, for all their power, still can't control what the farmer produces or decides to store or sell.

The point of the above discussion is that the farmer should not be a speculator. He has neither the time, contacts, nor resources to compete with the professional participants. But what he does have is the knowledge of when and what he's going to plant, the status of his crops as they mature, and what price is necessary for him to make a reasonable profit. The professional market participant, even with all of his vast information resources, can't fully determine what you or any other farmer precisely know about your crops. So, you do have an edge.

Weather is one of the biggest ingredients in this "edge." Your knowledge of the effect of weather on the commodities of most importance to you, and of the latest forecasts, can be just as good as the professional participants. Also, you should have a better feel than anyone else for both the current and projected effect of weather on your land. Using the steps outlined in this chapter will keep you as well informed about weather as you possibly can be at the present time. This is vitally important because of the underlying and consistently significant effect of weather on the fundamental factors of supply and demand.

It is beyond the scope of this book and not the proper position for a consulting meteorologist to advise you about the many things you must do to stay abreast of all the many developments and factors which can affect the price you receive for your crops. You will need to research this for yourself by seeking out those who know the market and who have your interests at heart. However, it is proper for me to advise you about the applications of weather knowledge—i.e., weather management—to crop marketing. There are steps you can take today to use weather information to your advantage as you market your crops. Your unique information advantage, versus the professional participants, allows you always to have a solid position from which you can bargain. This position is directly affected by weather. As mentioned earlier, the effect of weather on your crops, multiplied millions of times to include all crops worldwide, should be the basic reference point from which you begin to use weather management in crop marketing. By keeping advised of the weather worldwide and its resultant effect on crops, you will have a much sounder idea of the current and future prospects for price.

APPLYING WEATHER MANAGEMENT TO CROP MARKETING

Hence, the very first step in applying weather management to crop marketing is to keep fully informed about the current status of worldwide weather and its effect on crop status. You cannot hope to keep abreast of daily and hourly developments worldwide, so don't try (at least not yet). But you can stay very well informed about any and all significant foreign weather developments on at least a weekly basis and domestic weather developments on a daily basis. These sources are readily available.

The single most effective (for now at least) and inexpensive means of obtaining good information about weekly developments in both foreign and domestic weather is the *Weekly Weather and Crop Bulletin,* a joint publication of the National Oceanic and Atmospheric Administration Environmental Data and Information Service and the U.S. Department of Agriculture (Economics, Statistics, and Cooperatives Service). This publication contains reports of weather conditions on a state-by-state basis for the U.S. and for the most significant foreign crop-producing areas. It also discusses crop status, planting and harvesting status, and other factors designed to keep you informed about the effect of weather on agriculture. This publication will keep you advised about the wet weather that interrupted the cotton harvest in the Southwest or the planting of soybeans in Brazil.

Keeping informed about daily weather developments may prove to be a little more difficult depending on the sources available to you in your area (see chapter two). You will have to investigate both commercial radio and television and the NOAA Weather Radio to see which one, or combination, can keep you best informed about daily developments across the continental U.S. One of the more acceptable sources for this type of weather information is the weather portion of the morning network programs out of New York. These weather programs are national in scope and are usually prepared by a meteorologist.

This first step in applying weather management to crop marketing is the easiest but is quite necessary. You'll have an awfully difficult time trying to assess the future trend of crop prices if you're not fully aware of the current status of crops around the world. An associate and fellow consulting meteorologist often reminds me that you must know where you've been and where you stand now before you can accurately assess where you're going. Too often the emphasis in meteorology is on forecasting, instead of the analysis of what has already occurred and what that means to you right now. So, this first step involves the analysis of actual fact—i.e., weather that has already occurred and its resultant effect on the status of crops (both short-term and long-term as discussed above).

For supplemental information to the *Crop Bulletin* and general marketing advisories, you should also subscribe to the *Farmers' Newsletter* and *Commodity Outlook,* both published by the U.S.D.A. (Economics, Statistics, and Cooperatives Service). The publications are free and are specifically directed at farmers. Both the *Newsletter* and the *Outlook* are designed to help you with your marketing and production decisions; there are separate editions for oilseeds, cotton, wheat, livestock, and feed, plus a general letter. These same people also prepare a recorded message which they call the "Farmers' Newsline;" it gives you the latest crop, livestock, and economic information and you may dial it toll free at 1-800-424-7964. You should also contact your nearest U.S.D.A. office to find out the number of a similar recording for local market news highlights which are similarly recorded on a daily basis.

It would also be an excellent idea for you to obtain a set of color-coded maps depicting the major crop-producing and livestock areas across the U.S. (and the world, if possible). Display these prominently in your office or workshop to remind yourself continually where you should be focusing most of your attention. Several magazines and commodities firms can mail you a free set of these maps. It would be a good idea to provide a set for your local television weather forecaster just to make sure he's aware of where you'd like him to focus his attention.

Another publication designed to keep you better informed is the *Weekly Weather/Crop Assessment* published by NOAA (Environmental Data Service). This report is more expensive and is directed solely at keeping you informed about foreign weather developments. You should consider this report as only a possible option for you, if you have the time and the money, but it would be an excellent publication for your broker or marketing consultant.

In your efforts to keep fully informed, which is the first step to weather management as applied to marketing, you should also investigate other alternatives. There is a growing effort on the part of many, associated with agribusiness, to provide more and better information to the farmer. This ranges from publications to sophisticated, computer-driven information systems. There is also a growing realization that farmers nationwide are thirsting for knowledge and more sophisticated means of coping with the extreme complexities in the marketing process. This has resulted in a wealth of seminars about every conceivable topic related to agriculture. You should investigate all of these sources and participate in as many as is economically feasible. You still won't be as informed as the professional participants in the commodities markets but you'll certainly be holding your own.

In the final analysis, then, the first step isn't all that difficult; it just requires that you spend a considerable amount of time staying as fully informed as you can about all possible factors which can affect the price you receive for your crops. Weather, though one of the most important ingredients, is still just one of many important factors that affect the commodity markets. You can very easily stay well informed about all the factors of possible importance, including weather. The next step isn't so easy but is perhaps even more important.

THE SECOND STEP

The second step in applying weather management to crop marketing is the most difficult and potentially treacherous. This involves the forecasting of crop prices as related to weather. Some of the examples discussed earlier in this chapter gave you an idea of how weather can affect crop prices on both a short-term and long-term basis. But they also pointed out that just because the weather does one thing does not always mean the market will react accordingly. There are many other factors involved.

The primary reason this step is so difficult is that it involves forecasting of weather and other factors plus their interrelationships. It's not that difficult to obtain reasonably accurate short- and long-range weather forecasts if you know where to look. It is also possible to obtain respectable outlooks for the other major factors which are most likely to affect the commodities markets. But putting them all together and trying to make sense out of them may be another matter.

There are many sources for obtaining information and outlooks for all factors, including weather, which are most likely to affect the price you receive for your crops. Once again, you would do well to subscribe to the *Weekly Weather and Crop Bulletin* which includes both monthly and seasonal (as applicable) weather outlooks for the continental U.S. These outlooks are accurate enough to provide you with at least a good educated guess as to what the weather (temperature and precipitation) is most likely to do during the respective time periods. You should supplement this with non-weather information provided in *Commodity Outlook* and the *Farmers' Newsletter,* also mentioned earlier in this chapter.

The above publications will help as you plan and make your marketing and production decisions. The best rule of thumb to follow is that you can never have too much information to assist you as you make these decisions. But, as explained earlier in this chapter, you will almost always be at a disadvantage in the information business because of the vast resources and capabilities of the professional participants in the commodities markets. It is, perhaps, one of the most information-intensive industries in the world and the better informed you are, the more successful you'll be.

If the odds are so stacked against you, how can you hope to compete? You will have to seek out professional consultants in both weather and marketing to assist you in keeping fully informed and prepared to make the right decisions at the right time. These people exist and there are several who are quite good at what they do. But they are going to charge you at least a reasonable fee for their services. In most cases, however, the fee is miniscule compared to the return on investment.

Marketing consultants, by necessity, must also keep informed about weather developments worldwide. Some do a very credible job. However, there are two reasons you should get your weather information firsthand: 1) you can never be sure of its accuracy when you hear it secondhand from a non-meteorologist and 2) it's not beyond the realm of possibility that this information could be biased or just completely false.

The unreliability of secondhand weather information should be assumed. If there are ten people in a room and the first person passes a rather involved message to the second and so on, the message is rarely intelligible by the time it gets to the tenth person. The farmer is usually the "tenth" person when it comes to weather information as it relates to the commodities markets.

Biased or false weather information is a fact of life when dealing with the commodity market. I have witnessed reports of hurricanes that didn't exist and frosts that had absolutely no chance of occurring. It is not necessary to elaborate but let it suffice to say that you must be extremely careful as you tread lightly about the market and make every effort to check the credibility of your sources.

With the above in mind we come to the point at which you must now ask yourself where to turn for consultation concerning both weather information and forecasts. The government publications noted above, especially the *Crop Bulletin,* are excellent sources. But, here again, you're dealing with yet another subject and a sheer volume of information that will almost certainly require outside help. Hence, you should seek out a competent consulting meteorologist to help you in applying weather management to crop marketing. See chapter three for information in helping you to pick a professional consulting meteorologist.

As with selecting a quality consultant who can provide you with marketing information and forecasts, you must be highly selective in choosing a consulting meteorologist. There are currently only a handful who can provide you with everything you need to know, including quality long-range forecasts tailored to your specific needs. You must be extremely careful, for there are a few unscrupulous ones who make fantastic claims of accuracy, often based on a single spectacularly successful forecast (among many bad ones), and will promise you anything just so long as you'll subscribe to their service. Use the forecast-accuracy statistics and discussion in chapters one and four to check on the accuracy of these consultants and ask them to back up their claims with fact.

As discussed in chapter four, it *is* possible to determine with respectable accuracy the probable course of weather for all major agricultural areas through an entire growing season. Only a very few meteorologists are capable of doing this at the present time. However, you should be aware that this capability does exist and make every effort to obtain forecasts of this nature for all major crop areas of interest to you. The current accuracy of these forecasts (from reputable sources) are sufficient to give you a good idea of the most likely trend of weather through the growing season. But they may get even better in the future because of the demand for forecasts of longer and longer length and as a result of better forecasting techniques coupled with ever more-sophisticated technological advances. The stakes are so fantastic that you can be sure of a continuing effort to improve long-range forecast accuracy.

For example, in March of 1978, one statistical outlook for the major producing areas indicated that an abnormally dry and hot summer was in store for the cotton-producing areas of the High Plains of Texas and portions of the Mid-South. This same outlook also called for near-normal weather conditions over the Main Corn and Bean Belt with the first killing freeze date to be near the average time of occurrence in just about all of the major producing areas for cotton, soybeans, corn, and rice. You should never gamble all your chips on any one forecast, but this particular forecast would have implied that, all other factors aside, cotton prices might well remain stable or rise higher and that corn and bean prices might decline.

Shortly after the above forecast was made, untimely rains and cold weather created horrible planting conditions for cotton in the Mississippi Delta, most of which had to be replanted, while it was too dry to plant in the High Plains of Texas. Cotton prices rose higher. That summer, severe drought hit Southeast Arkansas and the High Plains of Texas and cotton prices advanced again. Weather wasn't the only factor involved, or prices might well have soared because of a smaller crop, nearly one-fourth smaller than the preceding year; but mill demand at home was quite lackluster for the second year in a row. However, a strong foreign demand for U.S. cotton, in conjunction with the U.S. dollar's sagging position against the yen, kept prices stable and, as a result of a much smaller crop, allowed the price to rise. In this case, all factors, with the exception of weak U.S. mill demand, combined to raise prices.

In the Main Corn and Bean Belt, planting was a little late but it followed good rains in April and May. The crop quickly made up for lost time in ideal weather conditions and, because the first killing freeze in most locations was near or later than normal, a bumper crop was harvested. In this case one would have expected prices to sag considerably but exports were strong, as were domestic livestock markets and their need for soybean meal; and Brazil, the world's second largest producer and exporter of soybeans, was considerably crippled by drought. Also, farmers decided to store more of their crops than expected. In this case, weather information alone would have implied a pessimistic prospect for corn and soybean prices, but other factors prevailed.

Weather is undoubtedly one of the most significant and persistent factors affecting crop prices. It should be given considerable emphasis, tempered by other important factors as they arise. Hence, the best thing to do is to plan your marketing decisions each growing season using weather as a major input but in conjunction with other factors, such as foreign crop production. Since weather directly affects crop yields, long-range forecasts of the expected growing-season weather can be used as the basis for determining which direction yields are likely to go. This will give you your first estimate as to what and how much of each crop you should plant and how long you should wait before selling your crop for cash or hedging on the futures market. As the season progresses, you can keep track of the weather and how it's affecting crop status. Revised long-range forecasts will help you adjust your market position if necessary. As you get ready to sell, you can time your sell to coincide with a weather situation that will allow you to take advantage of a temporary rise in prices as a result of weather.

Thus, the first two steps in applying weather management to crop marketing involve 1) staying well informed about the current status of crops as a result of weather and 2) the projection of weather through the growing season in order to anticipate yields. As explained above, neither step is all that difficult to accomplish but the entire process can be made much simpler and more effective by consulting with a competent professional meteorologist who is familiar with the crop marketing process and your yields.

A POSSIBLE THIRD STEP

A third step, on the horizon but not yet a practical reality for most people, is the use of crop-yield models. This involves the construction of mathematical yield models to help give you a better estimate of your expected yields, and, hence, a better feel for when you should market your crop. A crop-yield model is constructed by showing the statistical association between critical weather variables and crop maturity and yields, in order to allow the user to assess the potential yield at any point in a growing season. This is based on the actual weather that has occurred up to that point. It will make a tremendous difference at the end of the year if you're getting paid for thirty—versus forty—bushel (per acre) soybeans. If you could more accurately determine this potential it would give you an edge in marketing your crops.

For example, if prices are high and you know that your yields will be low, you would do well to sell as much as you can now to protect your earnings. On the other hand, if prices are high and you know that your yields will be high, you could sell a portion of your crop as a hedge against lower prices and hold the remainder, providing that certain market factors are favorable, in hopes that prices will go higher.

Crop-yield models can be constructed for your area. It is a relatively simple statistical procedure to compare available yield and weather data for the area nearest you in order to construct a basic yield model which you can use as your first estimate. Of course, it won't be as accurate as it could be unless you've already got several years' worth of weather and crop data as prescribed in chapter five.

You can already make a pretty decent estimate of your crop's potential yield just by looking at it. But you could do much better than that by scientifically monitoring critical weather conditions (as discussed in chapter five) and, over a period of several years, begin to show a very good statistical association between weather and yields at practically any time during the growing season. This information, in conjunction with a long-range forecast through the end of the growing season, should provide you with a very nice estimate of expected yields.

For example, the basis for a yield model for cotton could be started from observing the association between the soil temperature at 4 inches and significant stages of maturity. It normally takes 100 hours of 60°F soil-temperature to germinate cotton to a stand. 600 hours of 60°F soil-temperatures are required from stand to first square and an additional 600 hours to first bloom. 2,000 hours of 60°F soil-temperatures are required from planting to mature and open boll. The above data applies specifically to certain varieties in the North Mississippi Delta but would be similar to other cotton areas and varieties. You could study this association with either a maximum/minimum soil thermometer or thermograph with soil sensor. (Both of these are shown in appendix B.)

The information above would be recorded in both the Daily Farm Weather Log and the Crop Maturity Log. When enough data is compiled, a mathematical model can be constructed showing the normal relationship between one or more weather variables and the various stages of crop maturity and yields. It is vitally important that accurate and consistent records be kept and, whenever possible, that yield data be given separately for each major growing area on your farm. The more data you collect, the more accurate your crop-yield model will be.

It is easy to see the potential use for accurate yield models but it is equally as easy to see that there's a long way to go before they're widely used and perfected. The use of mathematical yield models is by no means widespread or widely accepted but as more and more farmers begin keeping better weather and crop data there'll be more to work with and much better results.

However, this should not deter you from investigating several sources to determine if crop-yield models have already been prepared for your area or state. While there are no specific agencies that can be recommended, there are several private sources, most of whom are reputable, that can provide this information or the expertise with which to construct a crop-yield model. Check around but be awfully sure of your sources and rely on only the most credible. Try checking with a consulting meteorologist or someone involved with research for some of the larger commodities consulting firms.

In the final analysis it is possible to apply weather management to crop marketing. You should make every effort to do so because of the vitally important role weather plays in determining yields. You cannot hope to compete with professional participants in the commodities markets and shouldn't try. But you can stay well informed and participate at your own level by subscribing to various government and commercial publications and making use of sophisticated consultants who can save you a great deal of time and money despite their consulting fees. You should stay fully informed about worldwide weather and its effect on crop status and potential yields. Long- and short-range forecasts for the major crop-producing areas will help you to make a good educated guess about future price trends and short-term effects on supply, respectively. But don't rely on these forecasts as the last word, for their accuracy has limits and there are many other factors which also play an important role in determining crop prices.

You can and must stay fully informed about all facets of the crop-marketing process. By following the steps outlined in this chapter, you'll be as well informed as it is possible to be about the current and projected effect of weather on crops, worldwide. Weather is undoubtedly one of the single most important factors affecting the supply of and demand for crops. Your knowledge of this effect will allow you to time the sell of your crops intelligently and logically. This should securely and consistently place you in the top third of prices for your crop(s) in any growing season. Hence, the procedures outlined in at least the first two steps of this chapter should help you to start turning higher, more predictable profits immediately!

7
CONCLUSIONS

Weather is perhaps the most ignored, misunderstood, and seemingly capricious variable affecting the entire American economy, including agriculture. The whims of Mother Nature are weathered by individuals and businesses alike with stoical detachment and merely accepted as our lot in life or a business hazard that must simply be endured. This blissfully ignorant attitude about weather is literally costing our economy hundreds of billions of dollars annually and yet, incredibly, no one really seems to care. For it is just naturally assumed, and casually accepted by all, that absolutely nothing can be done about it.

This attitude is promoted by the flippant treatment of weather by the majority of radio and television stations and the national networks, by the enforced ignorance of most weather-dependent businesses which stupidly disregard huge losses resulting from weather yet make no attempt to rectify the situation, and by the professional community of meteorologists who, for the most part, complacently accept an undeserved credibility gap (largely created by radio and television), refuse to set meaningful standards, and prefer to keep their collective heads in the sand, helping no one and ignoring one of the most golden opportunities of which any scientific profession might conceive.

There is absolutely no excuse whatsoever for the larger radio and television stations not to employ, or at least consult with, an experienced meteorologist from that community. The usually humorous or very groovy announcers featured as weatherpersons and supposed experts may be entertaining but their lack of knowledge about weather and their nonsensical approach to weather broadcasts is costing you money.

It is a very frequent occurrence for any commercial meteorologist to receive phone calls from businessmen inquiring about the weather. But when told that there will be a fee for the meteorologist's forecasting or analysis services, the businessman usually loses interest very quickly. I personally have seen firms lose thousands of dollars in ruined concrete or asphalt, yet be completely flabbergasted when it is suggested that they consult with a meteorologist for several hundred dollars a year and quite possibly save over 100 times the cost of that service. They usually just assume that the meteorologist is no more accurate than the "experts" they listen to on radio and television.

Regretably, we meteorologists deserve the credibility gap that exists between our actual abilities and the public's opinion of our abilities. We have complacently allowed the communications media to abuse and misuse our information without the slightest effort on our part to rectify or improve the situation. As a supposed community of professionals we have set no consistent standards of quality and have sat mutely by while some of our more unscrupulous colleagues made absolute fools of us all. To compound the problem, we have even developed a convenient vocabulary of jargon with which to voice our forecasts which serves to confuse even further an already confused and skeptical public. And yet, in spite of ourselves, we face today a collective potential for our services that is little short of fantastic.

Modern commercial meteorology essentially dates from the post-World War II years, after the many fantastic improvements in aeronautical and atmospheric technology. Since that time, there have been a relative handful of meteorologists who have consulted with industrial and agricultural clients on a fee basis. Yet this idea has not flourished and has not been encouraged by our fellow professionals, most of whom are employed by the federal government, universities, and various industries, and who have neither the vision, ability, or intestinal fortitude to accept the challenge of private practice. However, industry and agriculture have increasingly become aware of the kind of results and dollar savings that can be obtained by consulting directly with experienced professional meteorologists. Hence, the horizon for commercial meteorology is slowly widening, and it won't be long before men of vision begin to step into this opening and begin expanding both the abilities and the scope of commercial meteorology to meet the weather needs of an ever more-educated and progressive clientele, who realize the fantastic potential for savings through credible weather information tailored to their specific needs.

The above was not meant to be a diatribe against either the communications media, industry, or my fellow professionals. The intent was for my caustic comments to be a catalyst for action, a means of awakening the reader to both the problem and the potential. For this book was designed to be a positive, optimistic, and upbeat vessel for rectifying a common problem and applying a specific set of skills to that problem for the benefit of the reader. After reading this book, you are hopefully more aware of the considerable scope of that problem but equally aware that solutions do exist and can be initiated immediately.

The message imparted by my book can be appreciated by anyone, even though the specific intent was to propose a course of action (called weather management) to the American farmer whereby he can begin to monitor, analyze, correct scientifically, and take advantage of the effects of weather on his farm. However, the techniques for weather management can be applied to any business that must deal with weather-related problems.

A construction company would be well advised to employ a weather network arrangement before and during any construction project. Highway developers, city planners, and development architects should be making thorough studies of temperature and precipitation profiles (as determined with isohyetal/isothermal maps) in order to chart and construct safer highways, expand or renovate sewage and drainage systems, and plan and develop large office, shopping, and housing developments, respectively. The general public can use the explanations of terminology and knowledge about inexpert radio and television weather broadcasters to help them plan their daily work or recreational activities or protect their loved ones from extreme weather conditions at home or on the road by learning how to interpret and tailor forecasts to their use, and to demand that radio and television provide only the very best weather information available and stop using the featherbrained, fast-talking announcers who don't know an isobar from a barbell!

To the farmer, this book should be a positive first step to correcting a problem that has plagued all farmers since the first crop was planted in the earth many thousands of years ago. Vast improvements have been made in farm technology, the use of chemicals and pesticides, land and soil management, crop production, and, consequently, yields. Yet weather continues to run more or less rampant through the agricultural world, slashing, attacking, vandalizing, pillaging, and destroying at will. Billions are spent on research and development of new farm machinery, new chemicals, and new crop varieties but comparatively nothing is spent on the research and development of methods to employ or enhance the weather management techniques described in this book. New irrigation equipment is appearing daily, yet most farmers who purchase this equipment haven't the slightest idea as to when to irrigate or how to coordinate a weather forecast with that decision.

If every farmer in America were to read and use the techniques of weather management described in this book it would literally mean a savings of several billion dollars a year. In fact, it is not too far-fetched to assert that several hundred million dollars could be saved each year if every farmer in America actually knew what "50% chance of rain" meant. Time and time again, I have witnessed or heard about farmers who have lost, and continue to lose, enormous amounts of money by applying pesticides that are washed off when the forecast "only" calls for a 30 or 40 or 50% chance of showers. That one bit of knowledge, described in this book, is worth a veritable fortune on its own.

It is, quite simply, incredible that our vast knowledge of and abilities in weather analysis and forecasting have not yet been universally applied to agriculture. It's not so much that the farmer is unwilling to listen but that no one has really bothered to enlighten him about what's available and how it can be used to his advantage. It is absolutely unbelievable that this virtually unlimited business potential has not yet been explored. I must continually pinch myself, shake my head, and inquire of the heavens how such a golden opportunity has been overlooked by so many. The ultimate beauty of this situation is that it is the perfect free enterprise arrangement in that both the farmer and the consulting meteorologist stand to gain immeasurably from their association.

Another message this book seeks to impart to the reader is that there is something you can do today to begin to rectify the situation immediately. For example, you should no longer stand for the literal garbage and slipshod weather information you are getting from weather announcers on radio and television who, for the most part, don't know any more about weather than most of us know about neurosurgery. I have frequently witnessed television weather announcers depicting a cold front nearly 400 miles from its actual location or calling for heavy showers to be moving in within the next hour, then stupidly showing a radar screen with all of the showers already 60 miles to the east and moving rapidly away. It is not at all uncommon to hear radio announcers reading a completely outdated forecast. You simply shouldn't stand for this, because it's costing you money to attempt to rely on weather information that's not worth the time the announcer takes to read or show it. Voice your opinion to radio and television and, at the same time, persuade the giant chemical and implement companies, who spend millions on radio and television advertising, to take their business to the stations that cooperate best.

Seek out professional meteorologists and encourage them to apply their skills to agriculture. Every minute you procrastinate, wondering whether or not we can actually do what we say, you're unnecessarily wasting both time and money.

See to it that universities and major agribusiness firms are conducting research on the effect of weather and teaching the principles of weather management to future farmers. It will take at least a decade even to make a major dent in the problem, so you need to get started as soon as possible.

Most important of all, begin to employ weather management on your farm. It won't be easy, but it's not supposed to be. The techniques outlined in chapters five and six should be just as much a part of your daily activities as surveying your crops, repairing your machinery, keeping your books, or marketing your crops. You shouldn't delay for a moment, for the more efficient you are, the better able you will be to handle whatever the future will bring, including the possibility of long stretches of very unfavorable growing weather.

The world is becoming more complex and interwoven all the time. You are much more vulnerable than most businessmen in the world community. Your lack of defense against the effect of international phenomena on the price of crops puts you at a distinct disadvantage. You need every equalizer you can get and weather management is one of the biggest steps you can take.

In my opinion, the introduction of weather management to farming could someday become as spectacular an achievement as the first flight of the Wright brothers. This obviously sounds presumptuous, but not when you consider how far we've yet to go and what we're likely to achieve. It never ceases to amaze me as to how many superior farming techniques are *not* being used. Considering what's *not* being done, the productivity of the American farmer is little short of fantastic. One of the best ways to start using more finesse, preparing for an uncertain future, and becoming a more effective farmer is to begin to manage the seemingly capricious effect of weather—perhaps the single greatest problem faced by all farmers everywhere—on your farm.

This book will be a complete success if only one farmer begins to employ every phase of weather management. Because, when he discovers the benefits, he will soon tell another farmer, who in turn will tell another, and another, and another . . .

APPENDIX A
APPLYING WEATHER MANAGEMENT TO YOUR FARM

The following is a complete list of recommendations to help you employ weather management on your farm. The checklist takes all of the guesswork out of knowing what you need to do in order to use all available weather information to your advantage. The object is to make you a more efficient and knowledgeable farmer, weatherwise, and in the process reduce your operating costs and increase your profits through better use of weather information.

The checklist is divided into two main parts with optional sections for important recommendations that would be beneficial but not essential. Each suggestion is referenced to a more detailed discussion of its merits in the text.

I. *The Application of Weather Management on Your Farm.* The following recommendations are based on the fact that you must know more about the specific effects of weather on your farm before you can attempt to manage them.

1. ________ Take aerial photographs of your land and construct a detailed topographical map (bas-relief if possible) of your farm, showing all significant features in and around your land. (Make a note of any major topographical features within an additional 25- to 50-mile radius.) Chapter five, p. 60.
2. ________ Install at least one base station on your farm to monitor the effects of weather. This is the very least you should do in order to initiate weather management on your farm.

Chapter five, pp. 60-61.

3. ________ Establish a rain-gauge network on your farm. Place gauges in and around each major growing area. Chapter five, pp. 61-68.
4. ________ Encourage your neighbors to install a weather network or at least a single rain gauge and maximum/minimum thermometer. Chapter five, p. 67.
5. ________ If you irrigate, begin checking for soil moisture and evapotranspiration and using this information plus weather forecasts to schedule your irrigation applications. Chapter five, pp. 68-69.
6. ________ Install evaporation pan near all irrigated fields and measure daily. Chapter five, p. 68.
7. ________ Measure soil moisture daily, or as needed, in all irrigated fields. Consult qualified agronomist to determine proper depth of soil-moisture probe or sample analysis. Chapter five, pp. 68-69.
8. ________ Place maximum/minimum thermometers at highest and lowest points of each major growing area. Chapter five, p. 70.
9. ________ Place weather vanes in each major growing area of your farm. Space similar to rain gauges. Construct airfoils or windbreaks where and when necessary. Chapter five, pp. 72-73.
10. ________ Keep accurate records of all weather data collected. Don't discard any records. If possible, have them periodically reviewed and analyzed by a consulting meteorologist. Chapter five, pp. 63-66, 73-77.
11. ________ Construct isohyetal/isothermal charts for precipitation and temperature to illustrate their distribution better. Isohyetal charts should be constructed after each rain system. Temperature charts for maximum and minimum temperatures should be prepared daily. Chapter five, pp. 63-66.
12. ________ Determine areas of greatest amount of cold air drainage and temperature variation. Compare crop yields in affected areas to more "normal" areas. Use this information to alter planting and seed variety. Chapter five, p. 70.
13. ________ Determine areas of highest sustained winds, particularly at those times when young seedlings are most vulnerable. Correlate this information with precipitation and temperature analyses to see if there is any specific and identifiable interaction. Construct windbreaks where necessary. Chapter

five, pp. 72-73.

14. ________ Construct a set of supplemental records consisting of a Crop Maturity Log, Farm Operations Log, and Five-Day Planning Log. Chapter five, pp. 73-77.
15. ________ Have a consulting meterologist construct a weather cycle analysis for your farm. Use this analysis, plus your Farm Weather Network data, to construct a five-year plan for your farm. Chapter five, p. 79.
16. ________ Determine what kind of weather information you're receiving from local radio, television, and written publications. If it's not what you need, tell them so and suggest ways to improve their weather information. Chapter two, pp. 29-31.
17. ________ Buy NOAA Weather Radio where this specialized government radio network exists. Chapter two, p. 31.
18. ________ Subscribe to *Weekly Weather and Crop Bulletin.* Chapter two, pp. 32-33.
19. ________ Obtain a copy of the *Local Climatology Data (LCD)* for the nearest Class A observing station of the National Weather Service. (Call or write the National Climatic Center in Asheville, N.C.) Chapter two, p. 33.
20. ________ Consult with a professional meteorologist. Chapter two, pp. 33-34.

II. *Weather Management in Crop Marketing.* The following recommendations are based on the fact that weather is one of the single most important factors affecting the prices you receive for your crops.

1. ________ Keep fully informed about worldwide weather and its effect on crop status. Chapter six, pp. 85-87.
2. ________ Learn how to forecast crop prices using weather as one of your forecasting tools. Chapter six, pp. 88-91.
3. ________ Subscribe to *Weekly Weather and Crop Bulletin* for worldwide weather and crop status. Chapter six, p. 86.
4. ________ Subscribe to the following free publications: *Farmers' Newsletter* (USDA), *Commodity Outlook* (USDA). Chapter six, p. 86-87.
5. ________ Purchase radio that will allow you to listen to weather broadcasts on NOAA Weather Radio. Chapter six, p. 86.
6. ________ Obtain copies of commodity maps showing distribution of major production areas for major crops and livestock from a major brokerage firm. Chapter six, p. 87.
7. ________ Use toll-free farmers' newsline (1-800-424-7964), Washington, D.C., for latest crop, livestock, and economic

information. Local market news highlights are also available from USDA in selected cities nationwide. Chapter six, p. 87.

OPTIONAL

8. ________ Construct crop-yield models for your crops and use them to help time your marketing decisions. Chapter six, pp. 91-93.
9. ________ Consult with a professional meteorologist with agricultural experience. Chapter six, p. 89.
10. ________ Subscribe to *Weekly Weather/Crop Assessment* published by the Environmental Data Service. Chapter six, p. 87.

APPENDIX B
WEATHER INSTRUMENTS

An accurate accounting of the effects of weather on your farm should be just as vital a function as keeping your books. The Internal Revenue Service frowns on farmers who don't keep good financial records; Mother Nature is no more tolerant of farmers who make no effort to keep records on local variations in the weather. Just as you do with your financial records, you must insure a strict cost accounting of weather, as discussed in chapter five.

There are two major reasons for installing a weather network on your farm. First of all, the only way you can accurately assess the effect of weather on your farm is to measure this effect physically with an assortment of instruments as described below. You may already have a pretty good idea of how weather is apparently affecting your farm. But until you actually quantify this effect you simply won't have the specific facts you need to make the critical management decisions related to weather (see chapter five for further details).

The second reason you should install a weather network is to provide the weather data essential for accurately tailoring a forecast to your farm. It is not unusual to have only one or two official weather observing sites per 10,000 square miles. Any additional information will help you and/or your consulting meteorologist to tailor forecasts to the most important piece of real estate in your world—your farm.

To measure the effect of weather on your farm you must use specialized instruments designed to sample the atmosphere or effects of weather near the surface of the earth. To accomplish this, there is a nearly infinite variety of gadgets, instruments, and indicators which respond to or measure individual elements of the atmosphere or weather. Of course, the larger the selection the harder it is for you to choose the right ones for you. That's what this Appendix is all about.

It seems rather incredible that just a few specialized instruments placed near the earth's surface can give you a relatively accurate representation of weather in the lower layer of the atmosphere (troposphere), which is actually several miles thick. Yet, your instruments will record the local variations of each major weather pattern and give you an excellent idea of what to expect whenever a high pressure cell floats overhead or a cold front pushes a line of thunderstorms in your direction. You will still be measuring the effect of all those miles of atmosphere overhead and of weather patterns hundreds of miles in diameter, but the local changes in the atmosphere and the weather as recorded by your instruments will reveal the patterns of weather peculiar to your area of which you need to be aware.

As with practically any measuring device, you generally get what you pay for. But that's not to say that you must spend a fortune to install a weather network (see chapter five) on your farm. You can effectively monitor the weather with inexpensive instruments that do the job as well as expensive ones. The important thing is not that you get precise measurements but that you get as many reasonably accurate measurements from as many different locations as possible. A simple but effective Farm Weather Network can be installed on most 1,000-acre or less farms for less than $500.

Before you purchase any instruments you should consider all the aspects related to installing and maintaining a Farm Weather Network. First, you will need to optimize the cost to an acceptable level depending on how many instruments you need and what percentage of your annual gross revenue you care to expend. Secondly, you will need to consider the destructive force of weather, especially freezing weather, and the need for either inexpensive, expendable instruments that can be cheaply replaced or more expensive and durable instruments at higher costs. Thirdly, there is the problem of vandalism, either in the form of the theft or destruction of your weather instruments. (Weekend sharpshooters love rain gauges.) Fourthly, you should consider the convenience (or inconvenience) of reading the instruments.

The very first step is to make a survey of your land and indicate on a detailed map of your farm the exact locations where weather instruments are needed (see chapter five for a detailed discussion of this procedure). Even if you're not going to construct a dense network, as recommended in chapter five, consider at least one set of instruments per major growing area on your farm. Once you've calculated the total number of each type of instrument you'll need, consider the location, cost, and protection of these instruments. If, for example, you calculate that you'll need 20 rain

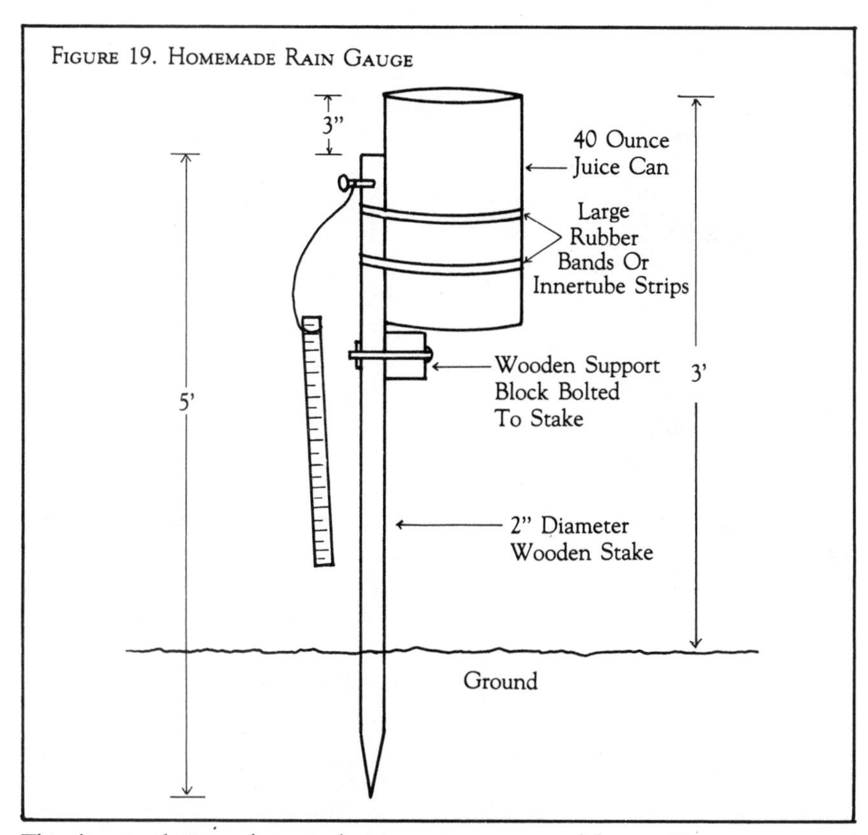

This diagram depicts a homemade rain gauge constructed from a 40-ounce juice can, a 5-foot wooden stake two inches in diameter, two large rubber bands, a two-inch wooden support block, a five-inch bolt, and a twelve-inch measuring stick.

gauges to construct a Farm Weather Network properly, select the very cheapest gauges for 19 of those gauges but purchase a more expensive and durable model to go in a protected area near your farm headquarters or home (whichever is closer to the actual growing areas).

Rain gauges are the most susceptible to weather damage and vandalism. I recommend a simple juice can, as shown in figure 19, which can tolerate freezing weather, will survive anything but a direct bull's eye at the base, is easy to clean and extremely inexpensive to replace. When in doubt, sacrifice quality for quantity; the more readings you take from as many different locations as possible, the more meaningful your data will be.

Ideally, you should distribute a complete Farm Weather Network as recommended in chapter five. However, since this will be a somewhat involved project, you will probably need to select the locations for the instruments you use, so that they are as close to your major growing areas as possible, yet still conveniently situated for easy reading. The larger the farm, the more difficult it will be to read several sets of weather instruments miles apart and in locations very difficult to get to, particularly after a rain. Hence, you might want to consider locating two or three of the expensive remotely readable instruments in the most important areas of your farm.

While the juice can is excellent for general distribution in your fields, I recommend at least one of the more expensive rain gauges for the most protected and secure location you can find nearest your home or farm headquarters. This will be the gauge that you always read, no matter what, at the proper time (see chapter five) following or during each round of precipitation. The expensive gauge will be the standard for all the others; it will be the most durable and accurate of all your gauges.

A similar logic should be followed in the selection of all other instruments. Consider the cheaper models for field distribution but select one of the more expensive models for the base station (see chapter five) in the most protected and secure location nearest your home or farm headquarters.

The easy part will be installing the instruments in their respective locations around your farm. The hard part will be reading the various gauges and/or emptying your rain gauges 365 days a year, year in and year out. It's as important as keeping accurate financial records but that doesn't make it any easier. To help ease the pain, you'll find handy sample recording forms immediately following the descriptions of the various instruments.

PRECIPITATION

The single most important variable you can record is precipitation. The more data you collect from as many different locations as possible, the better. Precision is not that critical, but always select a gauge from one similar to any of those shown which will give reasonably accurate readings for amounts of .05 inch or greater. *Do not* use the very small rain gauges often distributed as promotional items by many companies; they aren't accurate enough for your needs.

Attempt to mount all gauges as near as possible to official standards. The mouth or top of the gauge should be 3 feet above the earth's surface. Use only straight or right-angled cylinders to collect rain (don't use a bucket or similar container with sloping sides). Place gauge as far as possible from trees, fences, houses, buildings, or any other local obstructions which could significantly affect rainfall into the gauge.

This is not a complete list but shows an assortment of gauges ranging from very inexpensive ones you make yourself to sophisticated, expensive gauges that can be directly linked to a computer/communications system for instant recording, analysis, and transmission.

Least expensive

1. Homemade Rain Gauge—Purchase in supermarket or get used 40-ounce juice cans from neighbors, restaurants, schools, etc. This is least expensive yet durable, reasonably accurate, easy to maintain, and easy to read. Mount on 5-foot by 2-inch stake as shown. Can be used for both rain and snow. (See figure 19.)
2. Simple Yardstick—Used for snow measurements. Push yardstick through snow to top of ground and read depth. Take at least three different readings several feet apart and away from all obstructions and/or snow drifts. The average of these readings will be the amount of snow you will record. Record both inches of snow and its liquid equivalent. The general rule of thumb is that one inch of snow is equal to one-tenth inch of rain. In other words, divide the amount of snow by ten to get its liquid precipitation equivalent.

More expensive

3. Official Rain and Snow Gauge—This is the standard in all meteorological and climatological stations in the U.S. It would be an excellent choice for your base station(s). It is very durable, very accurate, easy to maintain, and easy to install. Order from specialized weather instrument manufacturer. Can be used for both rain and snow. (See figure 20.)
4. Forrester Rain Gauge—Meets official standards and is durable, accurate, easy to maintain and install. Can be used for both rain and snow. A good choice for your base station(s). (See figure 21.)

Expensive

5. Remote-Reading Rain Gauge—An accurate gauge that can be placed anywhere on your farm while you can read the amount of rainfall on a resettable counter in your home or office. An opening in the gauge allows precipitation to fall to the ground so that you never have to empty it. A very convenient, but expensive, gauge. A possible choice for your base station(s) but not advisable for general field distribution. Data can be fed directly to a computer. (See figures 22 and 23.)
6. Tipping-Bucket Rain Gauge—An accurate gauge which meets official specifications and also indicates rate of fall. Data can be transmitted to an assortment of remote readouts. This unit will

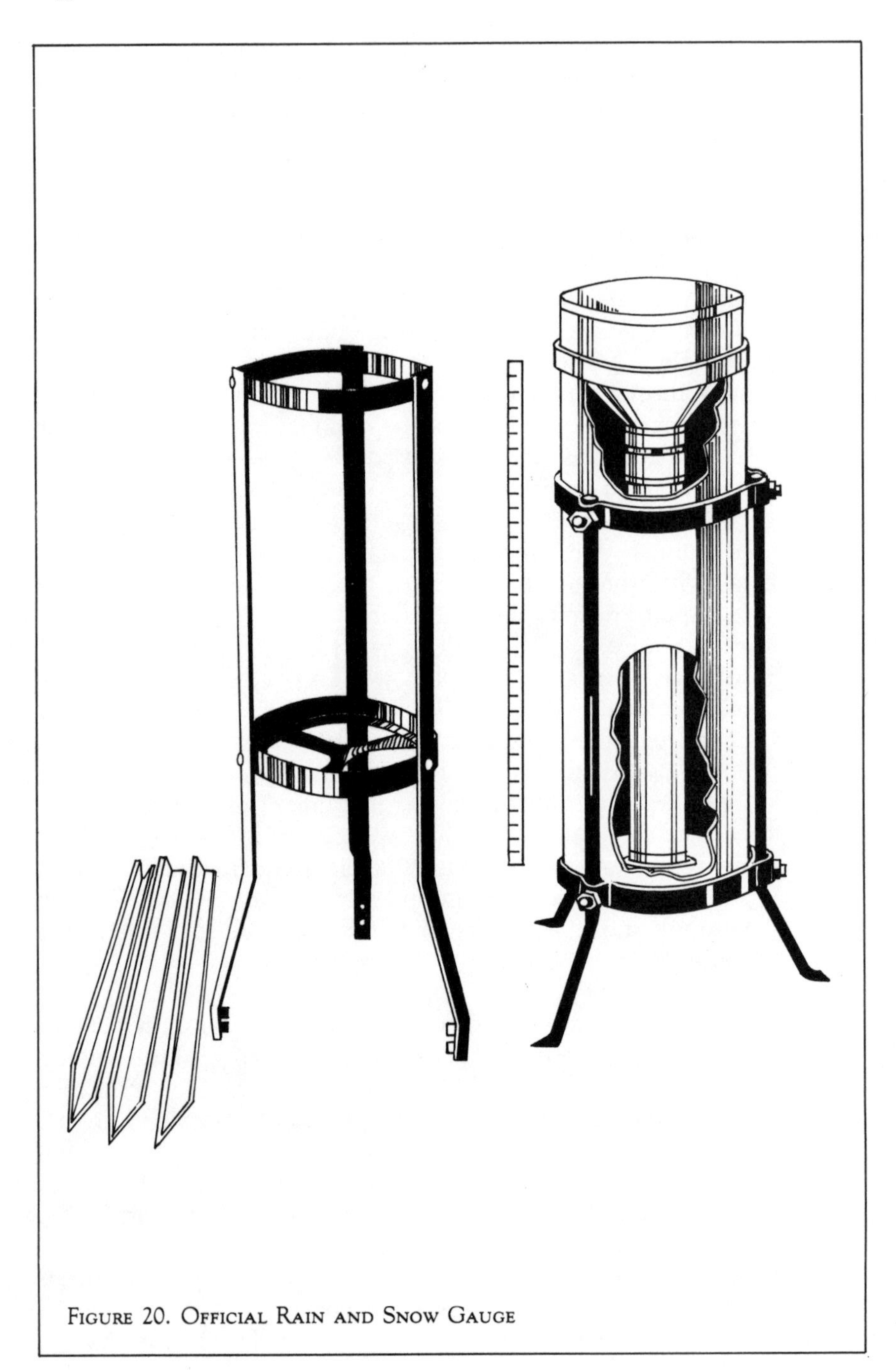

Figure 20. Official Rain and Snow Gauge

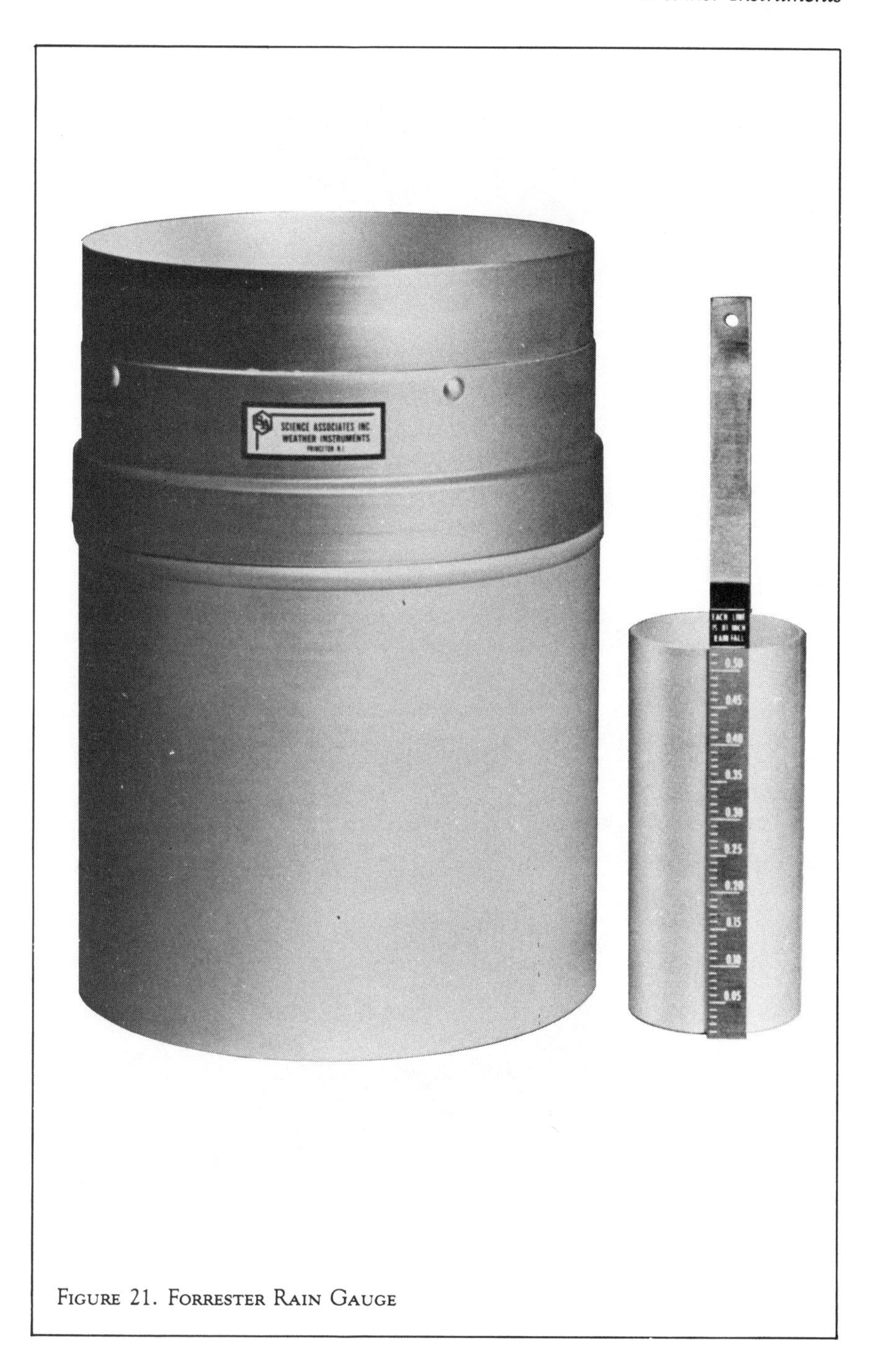

Figure 21. Forrester Rain Gauge

FIGURE 22. REMOTE-READING RAIN GAUGE

FIGURE 23. RESETTABLE COUNTER

Figure 24. Remote-Transmitting Rain and Snow Gauge

not function in below-freezing weather unless provided with a heating unit. It would be a good choice for a base station. Data can be fed directly to a computer.

7. Remote-Transmitting Rain and Snow Gauge—Different construction from tipping gauge. Meets official specifications and is accurate for amounts as small as .01 inch. A good choice for a base station. Can be fed directly to a computer. (See figure 24.)
8. Digital-Recording Precipitation Gauge—For rain and snow. Will encode rainfall data on punched tape or can be remotely fed directly to a computer. Can be left unattended for as long as five months. Has a capacity of 20 inches. Not the best choice for a base station because not as accurate as others shown above. This would be a good choice for remote locations hard to get to. (See figure 25.)

EVAPORATION

Irrigation scheduling is tricky business, so the more information you have, the better. Even if you're aware of the amount of precipitation and the soil moisture, it would be still more helpful to have one additional measure of the supply or loss of moisture available to your crops. A simple evaporation pan as shown in figure 26 can give you one more check on the rate of loss of water vapor or moisture from the soil (this process is called evapotranspiration) around your crop.

1. Evaporation Station—Figure 26 shows a complete climatological and evaporation station with an evaporation pan in the center, an instrument shelter to its left, and to its right an Official Rain and Snow Gauge, with a Remote-Transmitting Rain and Snow Gauge farther to the right. The evaporation station itself consists of the evaporation pan on the platform at the earth's surface and inside it the hook gauge (figure 27) from which the amount of evaporation is read. The hook gauge rests on top of the stilling well (figure 28).

TEMPERATURE

The second most important variable you should monitor (depending on needs, location, and time of year) is temperature. This variable is perhaps the single best indicator of overall weather conditions on any given day because temperature is dependent on wind, clouds, and precipitation. By keeping a daily record of the maximum and minimum temperatures you will be constructing a good source for analyzing how your location fits into the long-term weather cycles for your general area. When used in conjunction with your precipitation data, you'll be getting

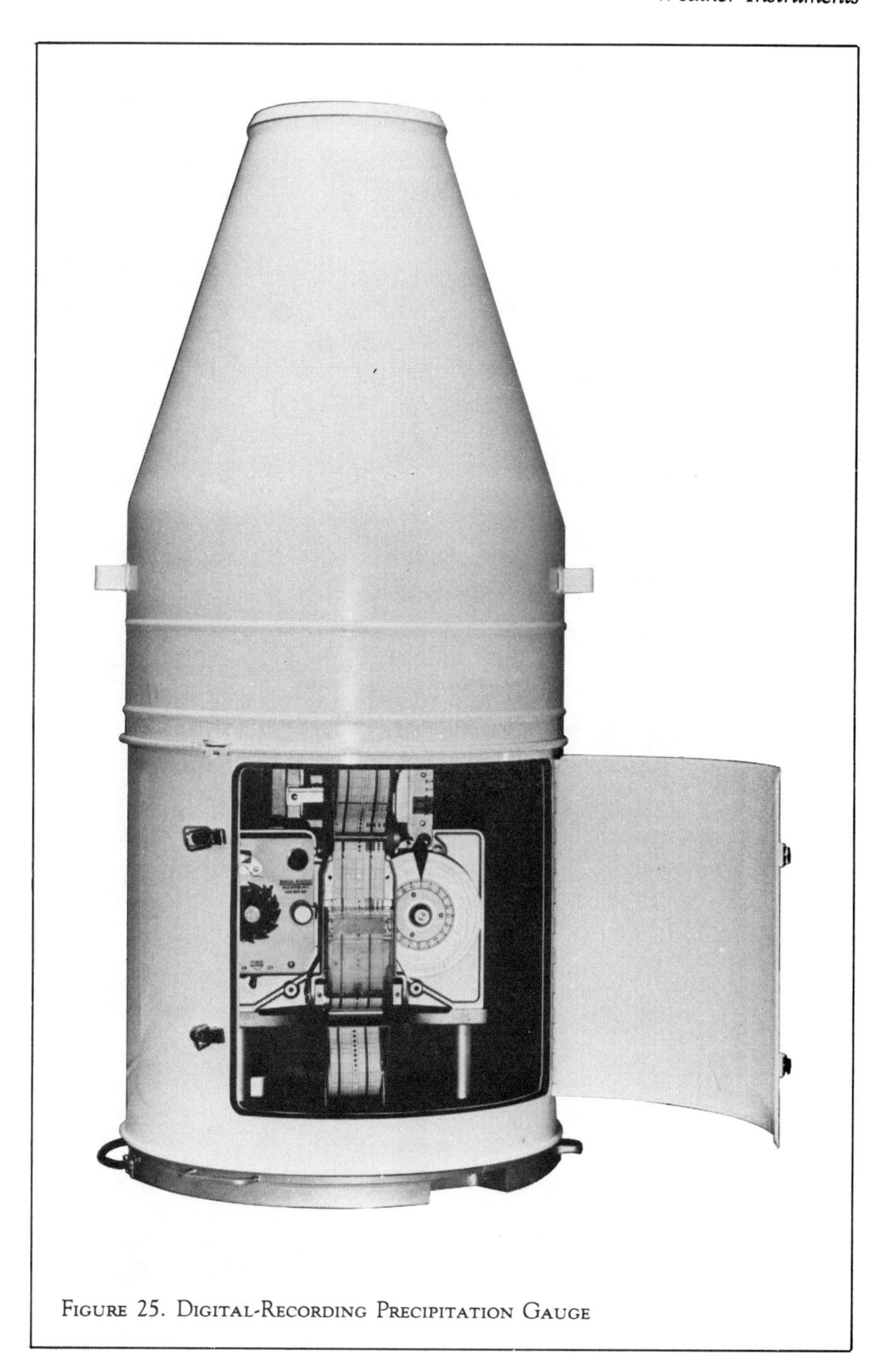

FIGURE 25. DIGITAL-RECORDING PRECIPITATION GAUGE

FIGURE 26. CLIMATOLOGICAL AND EVAPORATION STATION

Figure 27. Evaporation Station Hook Gauge

Figure 28. Evaporation Station Stilling Well

a very good start in quantifying the effect of weather on your farm.

Ideally, you should collect temperature readings from the lowest and highest points within your major growing area (see chapter five for a more detailed discussion). However, at the very least, you should have one good set of mercurial thermometers to record the maximum, minimum, and current temperatures. The one good set of thermometers should go in your base station.

When installing your thermometers, be careful not to mount them where there will be direct sunlight. If possible, place the thermometers in a shield or instrument shelter. Be sure not to mount the thermometers too low or too high; ideally they should be mounted as close as possible to the official height of 5 feet.

Here again, this is not intended to be a complete list of thermometers but just a representative sample of what's available from various specialized weather instrument manufacturers and, occasionally, various local stores. Soil thermometers are also included because of the importance of at least seasonal readings (as explained in chapter five).

Inexpensive

1. Maximum/Minimum Thermometer—A combination maximum/minimum thermometer which also indicates the current temperature. It is frequently called Six's thermometer after the original designer. Can be conveniently reset with a magnet. The range is from -40°F to +120°F. This would be a good choice for just about any use, even a base station, provided that it is properly shielded from the sun. (See figure 29.)
2. Etched Stem Maximum/Minimum Thermometer—Similar to no. 1 above, but a little more accurate and more conveniently mounted for field distribution when used in conjunction with the shield shown. The temperature range is from -40°F to +120°F. A magnet is used to reset the maximum and minimum indices. This would be a fine choice for a base station. (See figure 30.)

   The thermal shelter shown in figure 31 is recommended for agricultural applications and is made from a molded, highly reflective, impact-resistant thermoplastic. It is mounted facing north, on a post at a height of 5 feet above the ground.

Expensive

3. Maximum/Minimum Thermometers With Support—These thermometers meet official National Weather Service specifications and, when mounted on their accompanying support in an instrument shelter, constitute the most accurate temperature readings possible. There are more expensive and conve-

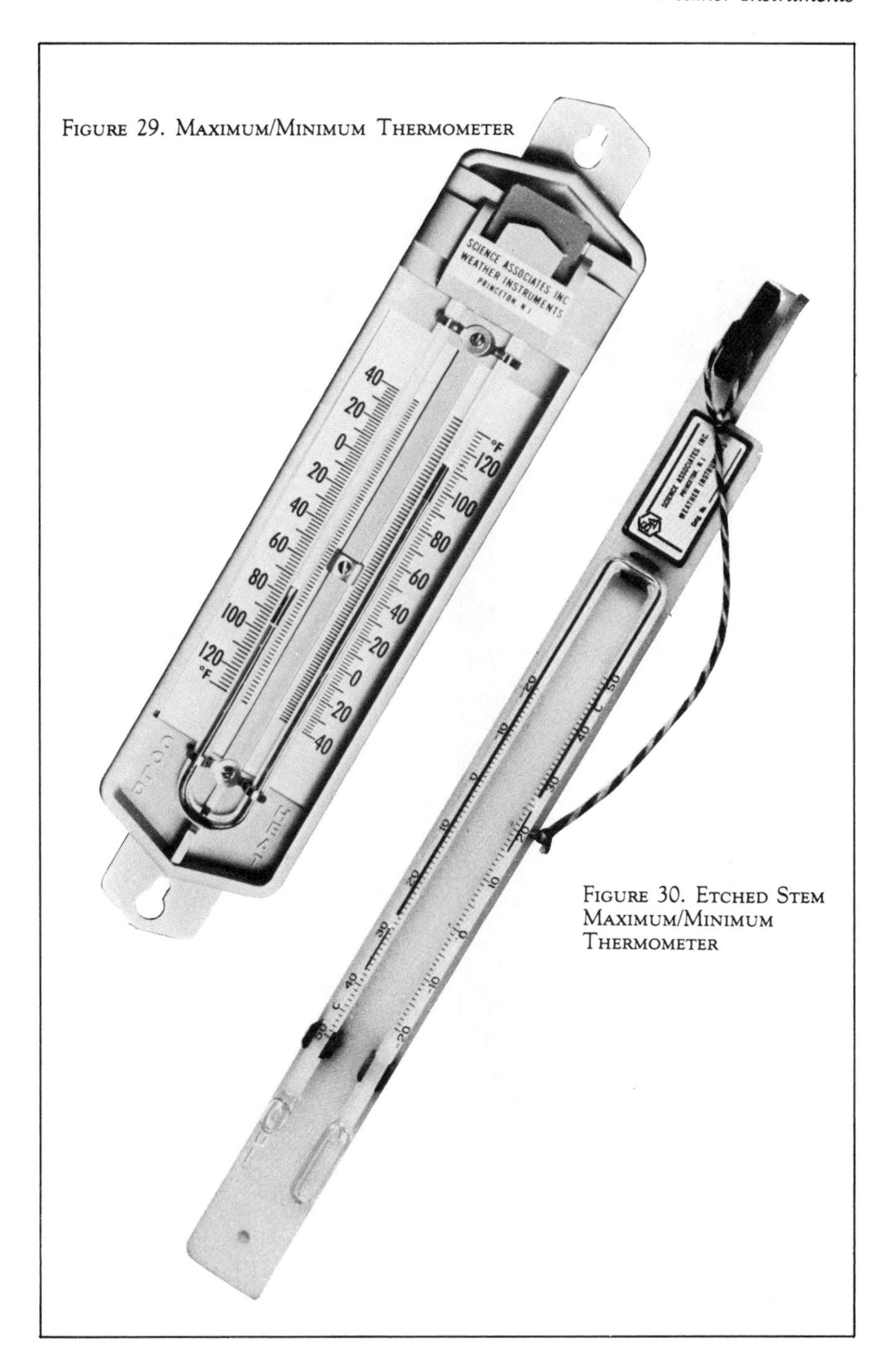

Figure 29. Maximum/Minimum Thermometer

Figure 30. Etched Stem Maximum/Minimum Thermometer

FIGURE 31. THERMAL-SHELTER, ORCHARD TYPE

nient systems available (as shown in 4) but they won't be any more accurate. The maximum thermometer is mercury-filled and the range is from -38°F to +130°F. The minimum thermometer is filled with amber spirits and has a range from -50°F to +120°F. The convenient support clamps each thermometer in correct position and allows them to be easily reset. The instrument shelter (in which these thermometers are normally mounted) is designed to protect the thermometers from direct sunshine, precipitation, and condensation, while allowing adequate ventilation. This would be an ideal arrangement for a base station. (See figures 32 and 33.)

4. Digital-Readout Thermometer—BCD Output (Binary-Coded-Decimal)—This thermometer is ideally suited for remote readings and integration into a complete data communications system. It won't be as accurate as No. 3, but it is far more convenient, and sufficient for all but highly accurate readings for research. The thermometer, housed in the sensor shield, is a linear thermistor. (See figures 34 and 35.)
5. Remote-Reading Recording Thermometers/Thermograph—There are many types but what they all do in common is to record the atmospheric and/or soil temperature on a strip of graph paper. The beauty of these instruments is that they can be left unattended for at least a week (even a month with some), yet you'll have a paper copy of the temperature at that location for the entire period. An instrument of this type might be a good choice for a remote location if the instrument could be properly secured and protected. (See figures 36, 37, 38, and 39.)

FIGURE 32. MAXIMUM/MINIMUM THERMOMETERS WITH SUPPORT

FIGURE 33. INSTRUMENT SHELTER

Figure 34. Thermometer Shield/Transmitter, Digital-Readout Thermometer Components/BCD

Figure 35. Readout Console, Digital-Readout Thermometer Component/BCD

FIGURE 36. REMOTE-READING RECORDING THERMOMETER

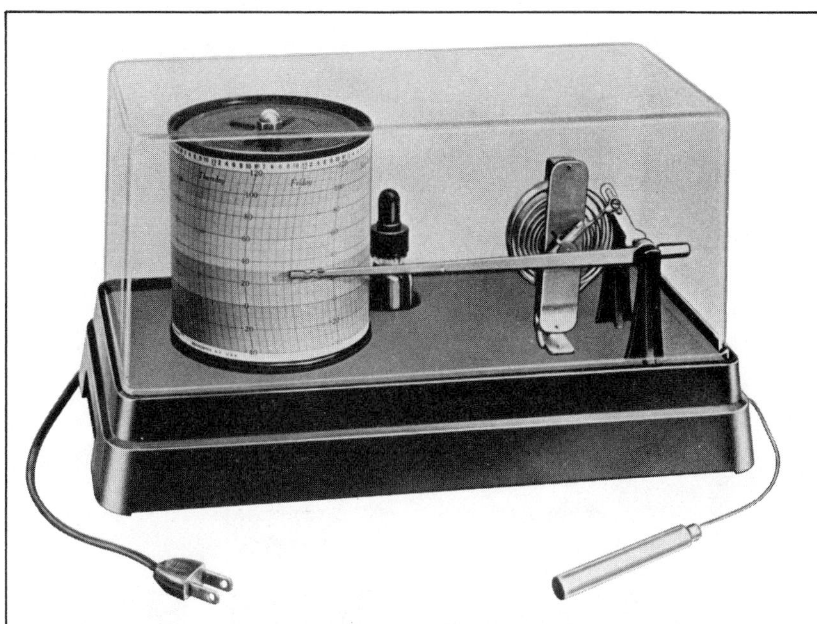

FIGURE 37. REMOTE-READING RECORDING THERMOMETER

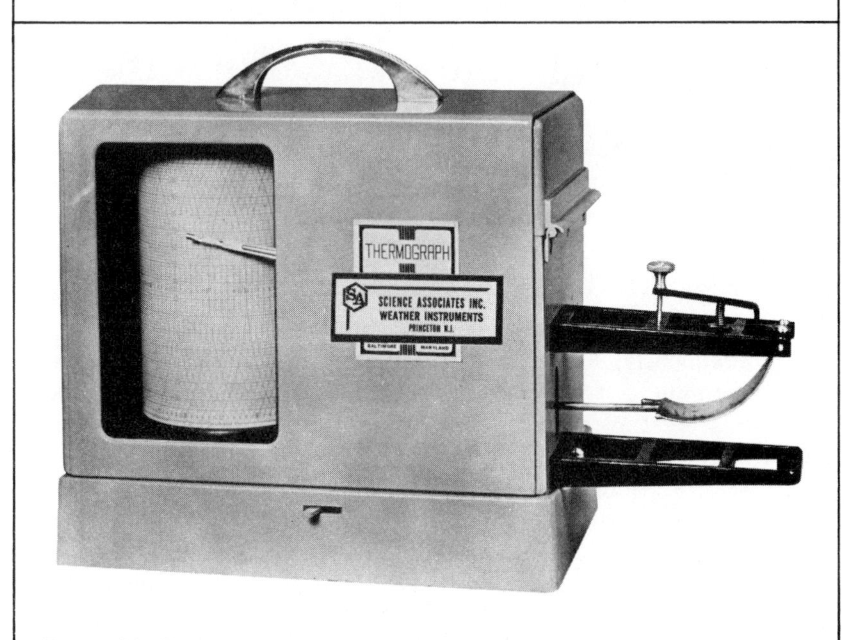

FIGURE 38. THERMOGRAPH

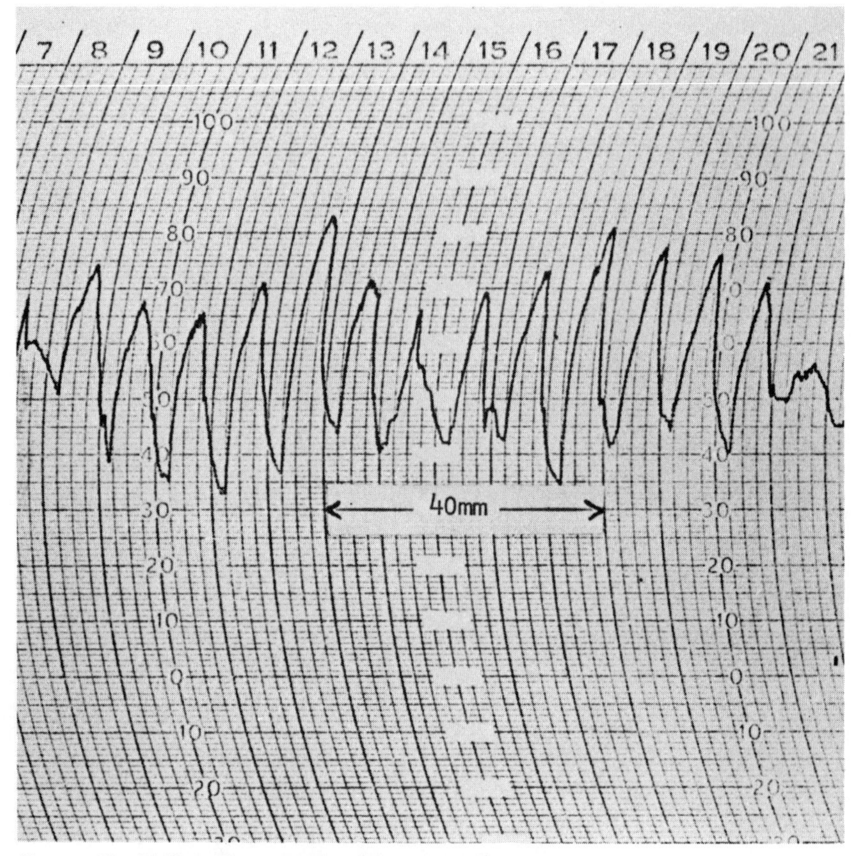

FIGURE 39. 31-DAY TEMPERATURE RECORDING CHART

SOIL TEMPERATURE

The soil temperature is vitally important to the maturity of your crop and, consequently, should be monitored. A more detailed discussion of this is given in chapter five. As with the temperature instruments above, there are several instruments that will do the job in measuring the soil temperature.

Inexpensive

1. Soil-Testing Mercury Thermometer—A mercurial thermometer conveniently mounted on a wood frame. Can be used for testing soil temperatures at a depth of about 1½ inches below the surface. (See figure 40.)
2. Soil-Testing Dial Thermometer—Thermometer is a helix bimetal sensor. Made of stainless steel and completely water-

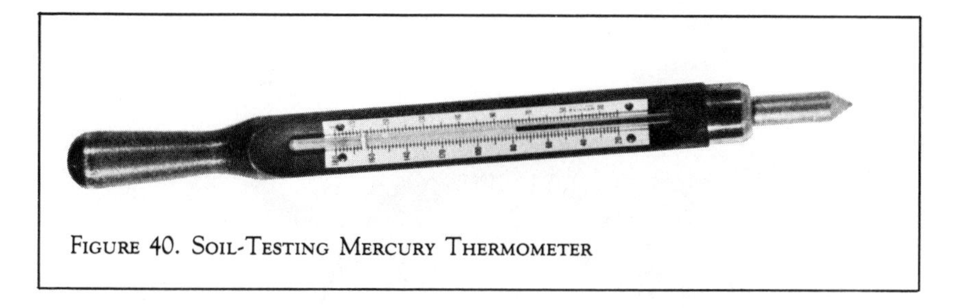

FIGURE 40. SOIL-TESTING MERCURY THERMOMETER

proof. It has a 6½-inch stem for measuring depths as much as 4 to 5 inches below the surface. (See figure 41.)

3. Soil Maximum/Minimum Thermometer—Uses an inflexible temperature-sensitive bulb 13 inches in length. The bulb is attached to the dial by a five-foot stainless-steel, flexible, spiral-armor, capillary tubing. This would make a good choice for a base station. (See figure 42.)

The third most important variable you should measure is open for debate. Wind is certainly important, but it is not an essential ingredient in establishing a usable set of statistics for a climatological analysis of your farm. However, the direction of the wind as well as its speed can be critical at several times during the growing season and a complete record of wind direction and speed for at least a single daily reading year-round can be a definite plus. Hence, wind instruments will be shown next.

WIND

To have a complete record of wind, you should indicate both direction and speed. It is sufficient to record this data once daily, preferably around 6 to 8 a.m., but one or two additional daily readings would be helpful, particularly after a marked change in direction. There will be a definite pattern of wind direction and speed that will prevail for each season of the year and in several different locations on your farm, particularly where the topography or amount of trees is considerably different from other places. A more detailed discussion of the phenomenon of wind and its significance to your farm is given in chapter five. You will need two instruments to accomplish the job: a wind vane (for measuring wind direction) and an anemometer (for measuring speed).

Inexpensive

1. Homemade Wind Vane—You can make this yourself with an old license plate, some aluminum conduits, and a little extra

FIGURE 41. SOIL-TESTING DIAL THERMOMETER

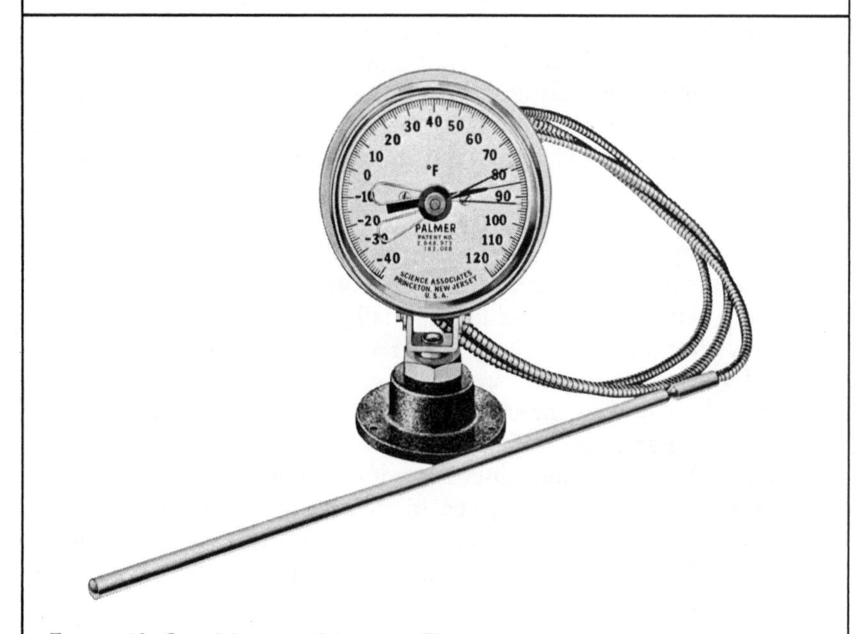

FIGURE 42. SOIL MAXIMUM/MINIMUM THERMOMETER

hardware. This would be an excellent vane for field distribution or just about any location (including a base station). It should be supplemented by no. 2, the Hand Windmeter, for determining speed.

Cut a slit in one end of four-foot aluminum conduit, insert license plate, drill two holes (as shown) and attach plate with bolts. Drill a larger hole in the middle (or at the balance point) and thread a 5-inch bolt through this hole. Put a washer on the bolt and position the wind vane on the aluminum conduit attached to the fence post. Make sure that the 5-inch bolt is

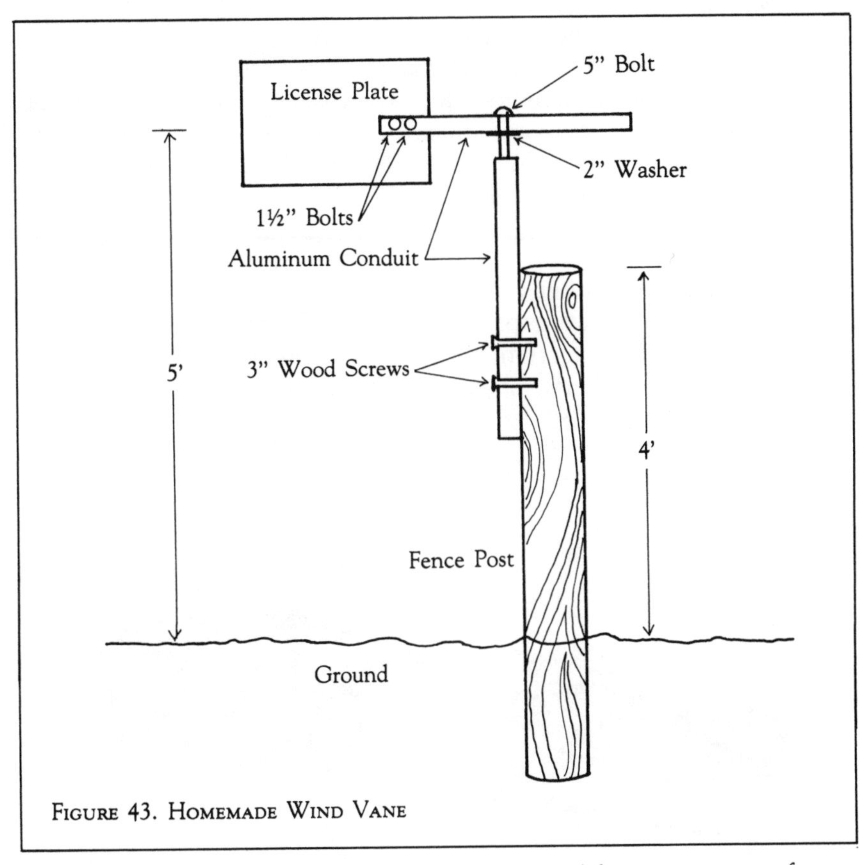

FIGURE 43. HOMEMADE WIND VANE

This diagram depicts a homemade wind vane constructed from two sections of one-inch diameter aluminum conduit four feet in length, an old license plate, two 1½-inch bolts, one five-inch bolt with two-inch aluminum washer, and two three-inch wood screws.

only slightly smaller in diameter than the upright aluminum conduit. Lubricate the bolt with heavy axle grease. (See figure 43.)

2. Hand Wind Meter—Accurate to within ½ m.p.h. on the low scale and 3 m.p.h. on the high scale. Hold back of meter in direction from which wind is blowing and read the speed on the appropriate dial. (See figure 44.)
3. Wind-Speed Gauge—A combination wind vane and pressure-type anemometer which can be conveniently mounted in the field or in your base station arrangement. (See figure 45.)

Very Expensive

4. Combination Wind Vane/Anemometer, Digital Readout, Direction and Speed—There are many types of expensive wind direction and speed instruments. This particular model is one of a set of instruments that also measures temperature, precipitation, pressure, and humidity. This set would be recommended for only base station locations because of the expense. (See figures 46 and 47.)

The last three weather variables to be discussed are relative humidity, soil moisture, and atmospheric pressure. The first two variables are important to crop production on your farm but the effect is usually seasonal. The atmospheric pressure is not an essential ingredient in a

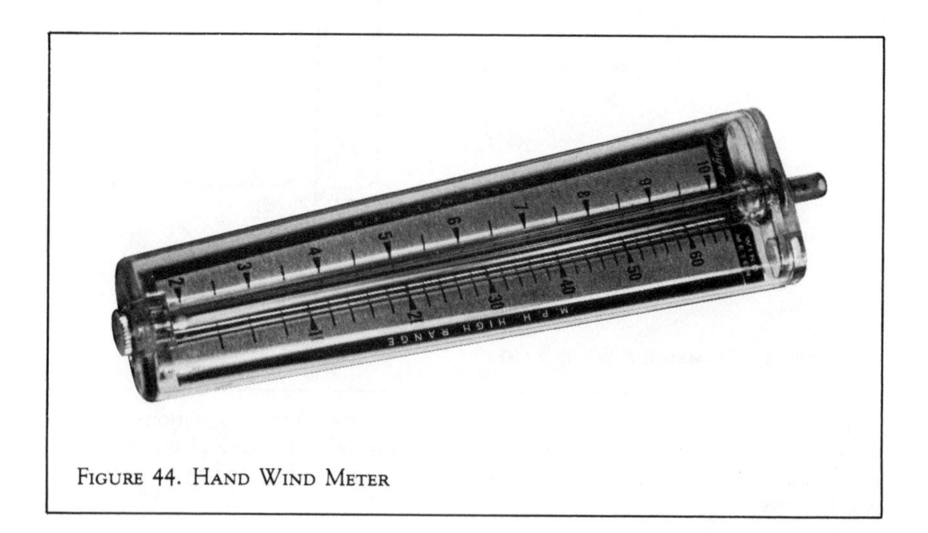

FIGURE 44. HAND WIND METER

Farm Weather Network but it can be useful as an indicator of expected weather patterns.

RELATIVE HUMIDITY

The most efficient and accurate method for measuring relative humidity is to use a sling psychrometer. It also happens to be the cheapest and most practical means for measuring relative humidity in several different locations around your farm. Hence, three inexpensive sling psychrometers are shown. A few additional and more expensive alternatives are shown, which offer greater convenience but no increase in accuracy.

Inexpensive

1. Official Sling Psychrometer—Meets National Weather Service standards for accuracy. Mercurial thermometers mounted on stainless steel backing. Purchase complete with wick and psychrometric table. (See figure 48.)

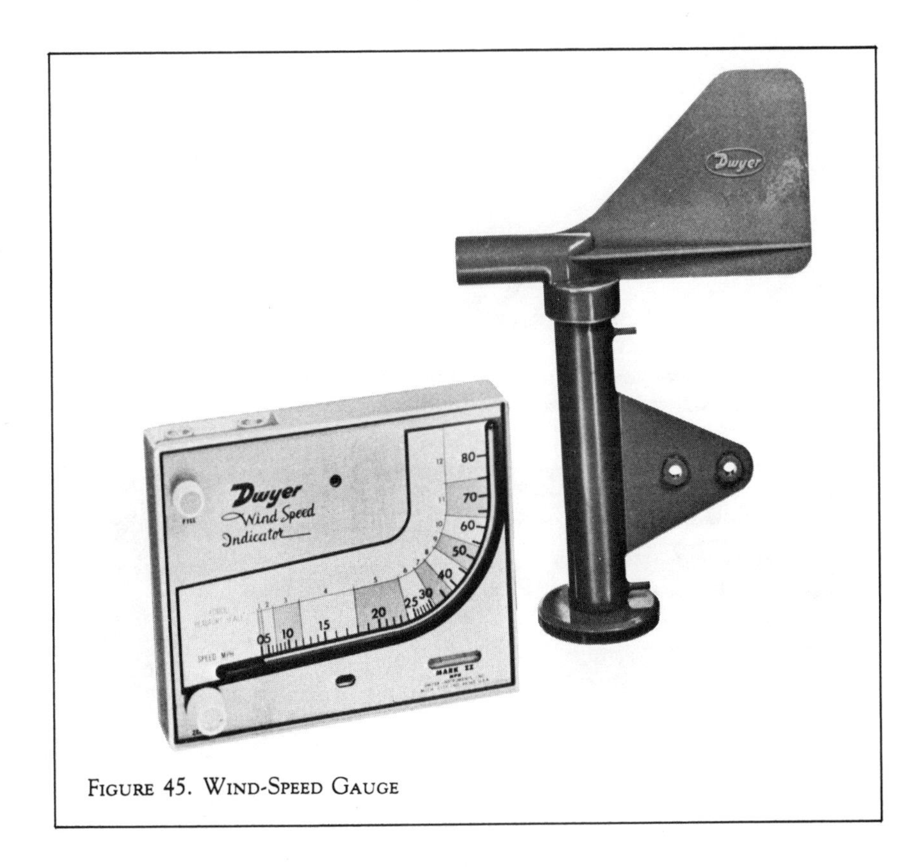

FIGURE 45. WIND-SPEED GAUGE

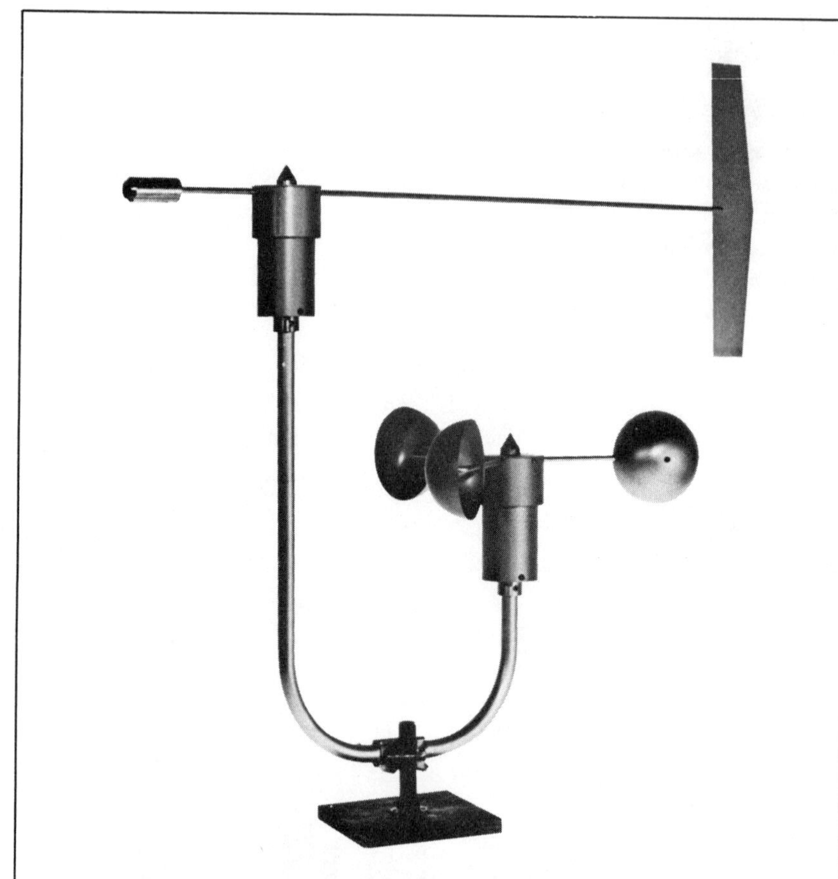

Figure 46. Combination Wind Vane/Anemometer, Components for Digital Readout, Direction and Speed

Figure 47. Readout Console for Digital Readout, Direction and Speed

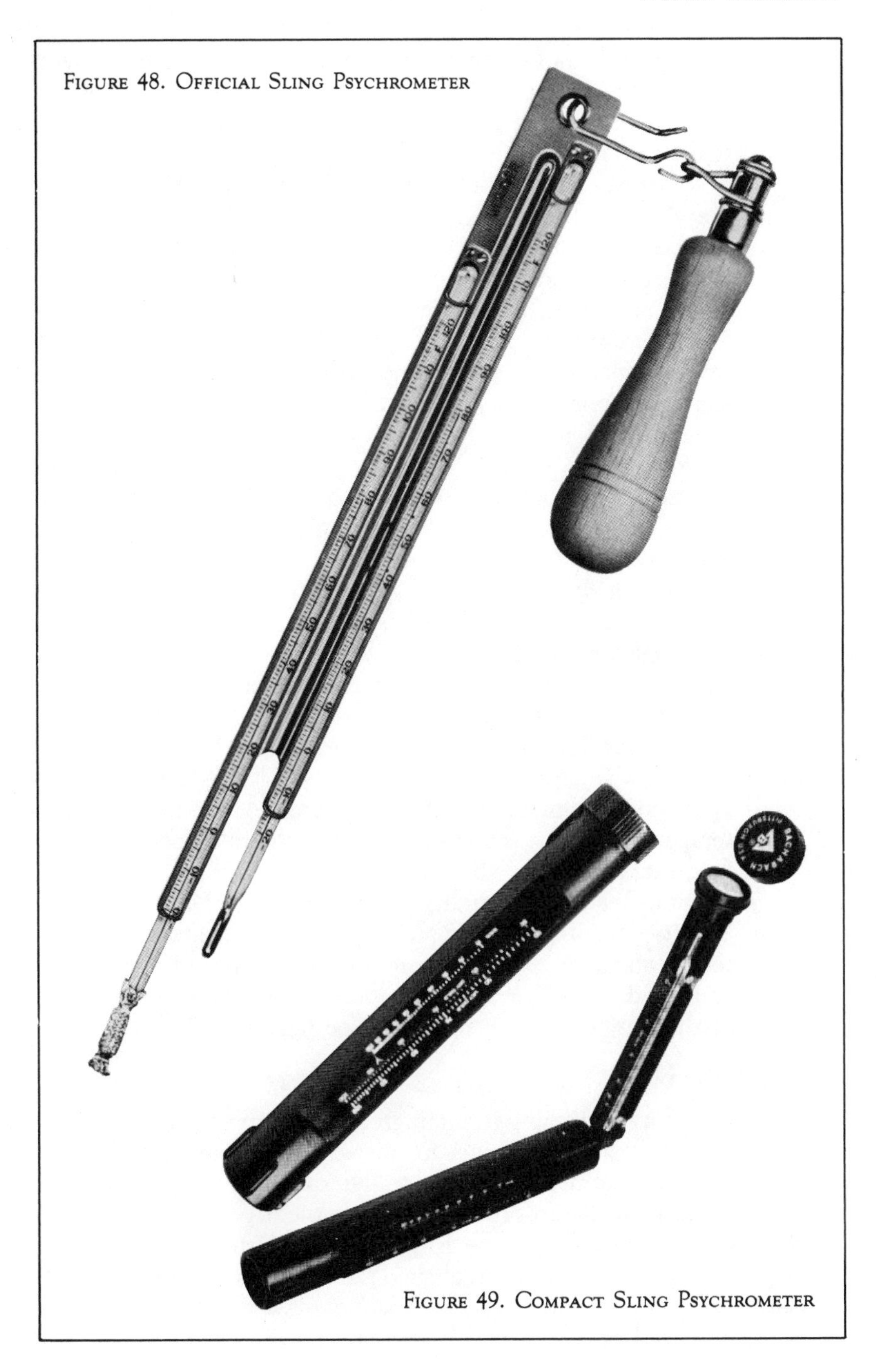

FIGURE 48. OFFICIAL SLING PSYCHROMETER

FIGURE 49. COMPACT SLING PSYCHROMETER

2. Compact Sling Psychrometer—Has a hinged thermometer case that slides into the handle for protection. Has built-in water reservoir that eliminates need to wet the wick. Also has built-in slide rule calculator for determining relative humidity from the wet and dry bulb readings. This would be an excellent portable psychrometer because of the many built-in conveniences. (See figure 49.)
3. Pocket Sling Psychrometer—A small, convenient pocket psychrometer with two 5-inch mercury thermometers. Thermometers are mounted on a flanged aluminum back, with swivel handle connection for whirling, and are carried in an aluminum-reinforced leather carrying case when not in use. (See figure 50.)

More Expensive

4. Psychron—A portable, battery-powered psychrometer with several advantages over sling-type devices: constant rate of ventilation; improved observational accuracy; less thermometer breakage; use in confined spaces; and illuminated scales for nighttime readings. (See figure 51.)
5. Direct-Reading Hygrometer—An easy-to-read, convenient instrument that uses an animal membrane element as a humidity sensor. This is highly portable and would be a good choice for field measurements. (See figure 52.)
6. Humidity/Temperature Indicator—A hand-held, battery-operated instrument for directly reading relative humidity and temperature. This is ideally suited for portable operations and field measurements. (See figure 53.)
7. Digital-Readout Humidity Indicator With BCD Output—This is one of a set of matching digital indicators for wind, temperature, humidity, pressure, and precipitation. This would be a good choice for a base station because of the convenience of the remote-readout feature. (See figures 54 and 55.)

SOIL MOISTURE

Soil moisture is a vitally important ingredient when scheduling irrigation. There are several techniques for measuring soil moisture but the most convenient technique appears to be the measurement by soil blocks. The soil block sensors can be conveniently placed at various depths and locations within the irrigated area. Determining the soil moisture is a simple matter of plugging the exposed leads into the measuring instrument. The measuring instrument is relatively expensive but you only need one. The soil blocks are quite inexpensive.

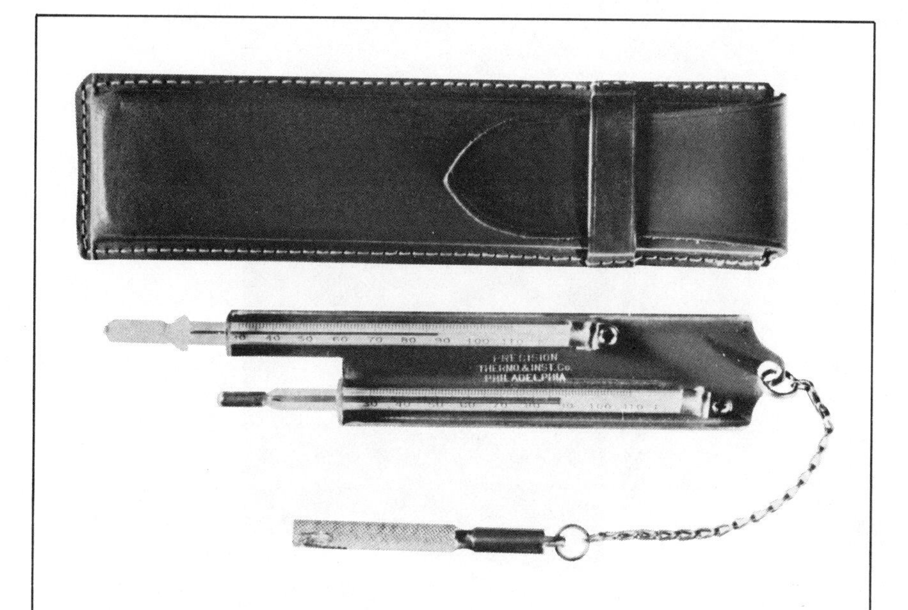

FIGURE 50. POCKET SLING PSYCHROMETER

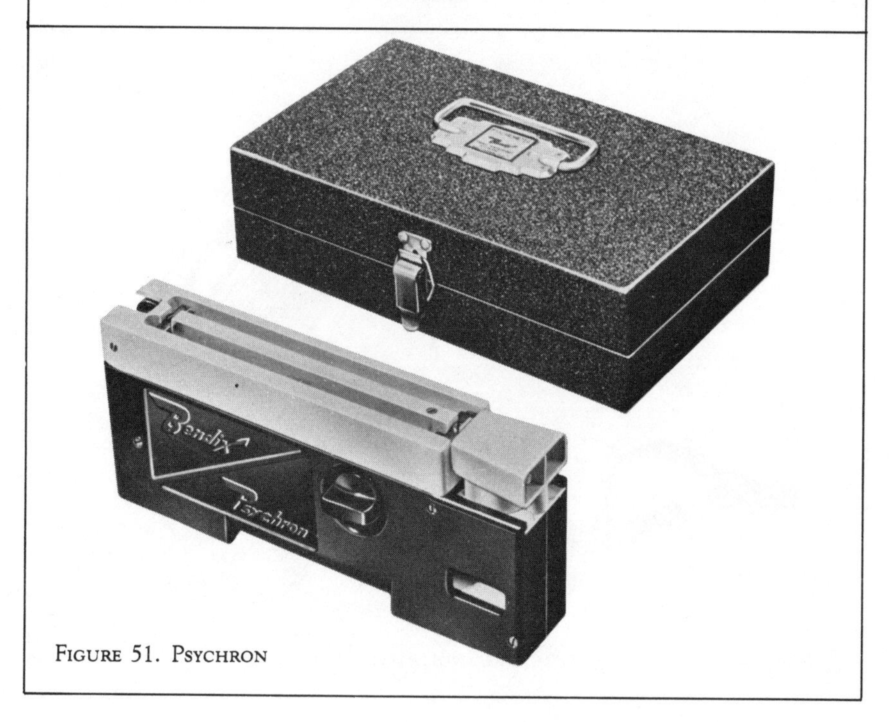

FIGURE 51. PSYCHRON

FIGURE 52. DIRECT-READING HYGROMETER

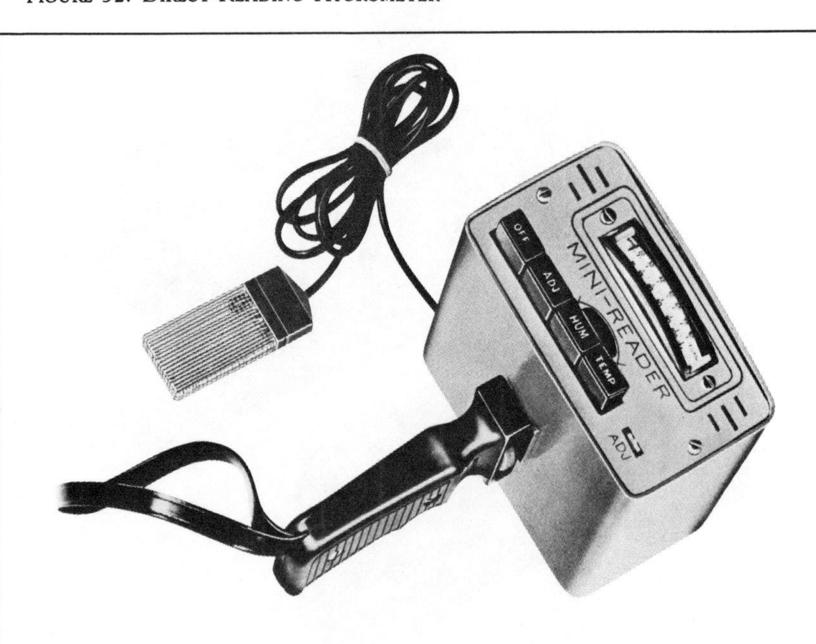

FIGURE 53. HUMIDITY/TEMPERATURE INDICATOR

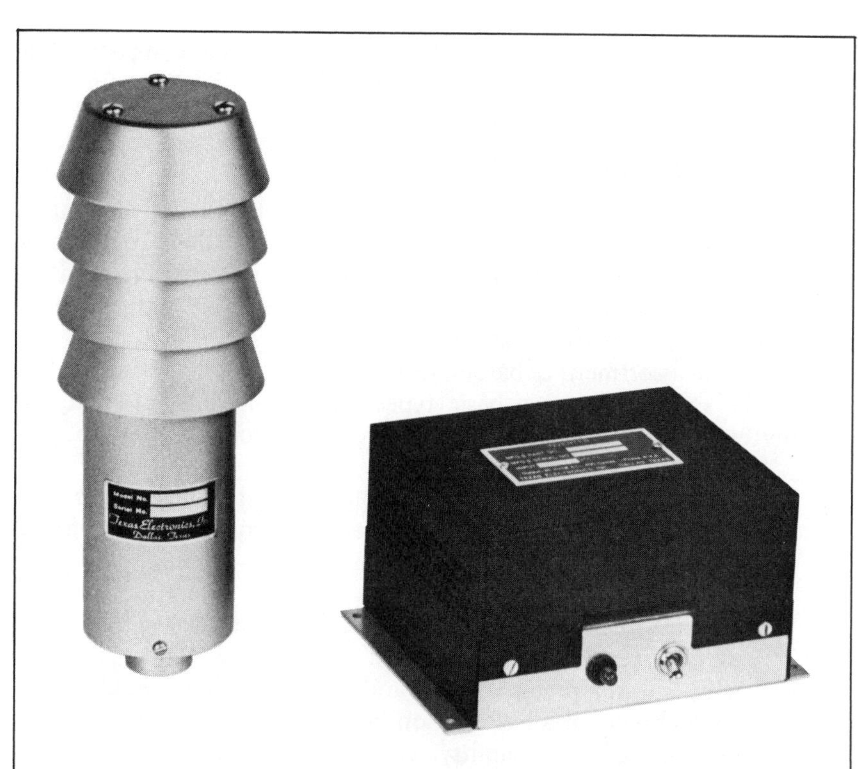

Figure 54. Hygrometer Shield/Transmitter for Digital-Readout Humidity Indicator with BCD Output

Figure 55. Readout Console for Digital-Readout Humidity Indicator with BCD Output

1. Delmhorst Soil-Moisture Instrument—A small compact solid-state tester. Operates on a 9-volt battery. Uses gypsum soil blocks as sensors. (See figure 56.)

ATMOSPHERIC PRESSURE

It is not absolutely essential that your weather network include instruments for measuring atmospheric pressure. However, daily readings of atmospheric pressure will be good supplemental information, and by carefully observing changes in the pressure you will be better able to anticipate changes in the weather.

Since the assortment of barometers is endless and fascinating, only a few will be shown. The two basic types are the aneroid barometer (a partially evacuated hollow disk) and mercurial barometer.

Inexpensive

1. Weather Barometer—An aneroid barometer. (See figure 57.)
2. Observer's Barometer—An aneroid barometer. (See figure 58.)

Very Expensive

3. Digital-Readout Barometer With BCD Output—This is one of a system of units designed to measure wind, pressure, temperature, precipitation, and humidity. This would be a good choice for a base station because of the convenience and remote-readout capability. (See figures 59 and 60.)

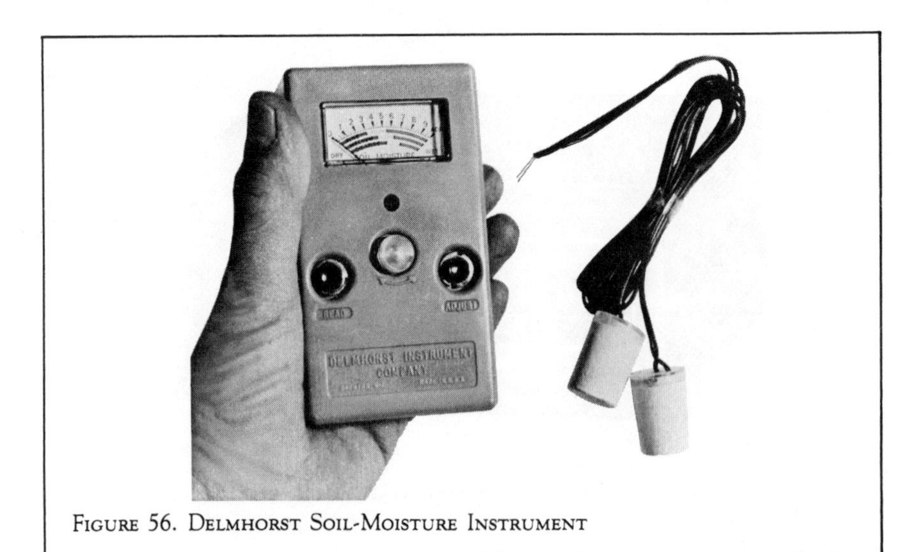

FIGURE 56. DELMHORST SOIL-MOISTURE INSTRUMENT

FIGURE 57. WEATHER BAROMETER

FIGURE 58. OBSERVER'S BAROMETER

Figure 59. Barometer Shield/Transmitter for Digital-Readout Barometer With BCD Output

Figure 60. Readout Console for Digital-Readout Barometer with BCD Output

Once you have selected the appropriate weather instruments and installed your Farm Weather Network, you must then begin to record all the different weather variables you'll monitor. The neater and more orderly the record, the greater its value. The first thing you must do is to indicate the precise location of each instrument on a detailed map of your farm. You should then prepare a smaller map (see figure 61) for encoding at least the daily precipitation (the single most important variable for preparing a climate analysis).

In figure 61, a blank circle and triangle were drawn at the location of each rain gauge and thermometer respectively. By preparing one master form and many additional copies you can conveniently enter the amount of precipitation and maximum or minimum temperature, in the appropriate blank, each time you check these instruments. When available, encode any additional variables (wind, soil moisture, soil temperature, relative humidity, pressure) that you have observed at or near any rain gauge or thermometer.

The master map in figure 61 will allow you to construct convenient isohyets (lines of equal amounts of precipitation for a given period of time, usually 24 hours, or a specific storm) as shown in chapter five. The object of these charts is to make the analysis of your field and base stations easier and more meaningful; you'll be able to see the pattern of precipitation at a glance. If your thermometer network is dense enough, you may also want to draw isotherms (lines of equal temperature). If you're going to combine both on the same chart, it would be better to use the common color scheme employed by meterologists: isohyets in green and isotherms in red. See chapter five for a more detailed discussion of isohyets and isotherms and their significance to your farming practices.

The isohyetal/isothermal map should be further supplemented by a daily log of current weather conditions as observed at your base station(s) plus a record of forecasts. A convenient form (the Daily Farm Weather Log in figure 62) is shown for encoding the most important variables, plus additional variables of seasonal or geographically isolated significance, the daily forecast from at least two local sources, additional comments of possible significance, and a reference to the corresponding isohyetal/isothermal map. You should keep a separate form for each base station.

When filling in the form (figure 62), you should be very careful to use a consistent scheme for logging the data and always enter the data in a waterproof ink. The current observation should be encoded at least once daily and more if possible. The maximum temperature will be the highest reading for the 24-hour period from midnight to midnight of that day and will usually occur in the afternoon. The minimum temperature is

FIGURE 61. ISOHYETAL/ISOTHERMAL MASTER MAP

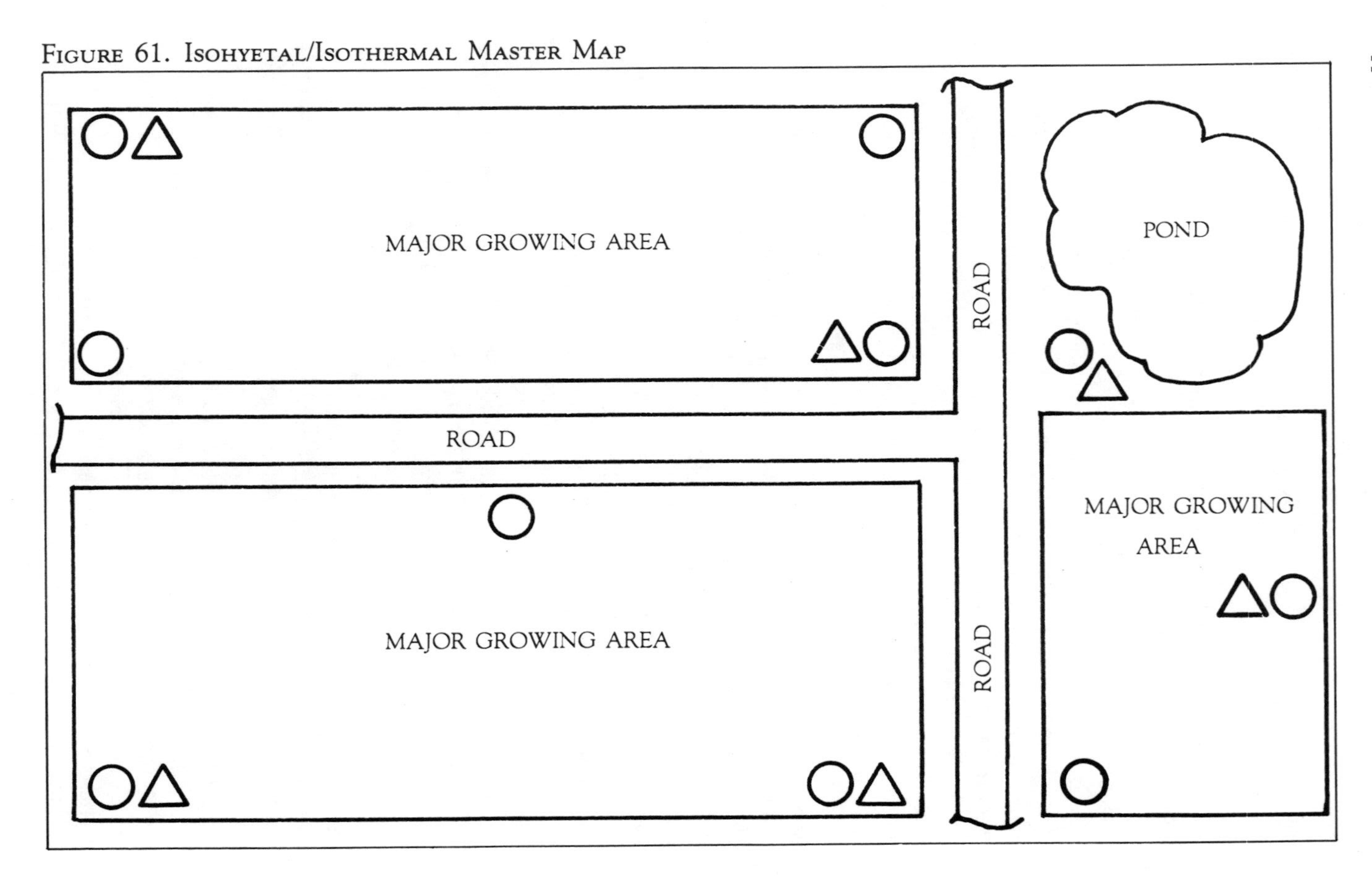

the lowest daily reading for the same period and usually occurs at dawn. Enter both wind speed (in m.p.h.) and direction (from which the wind is blowing). The pressure should be entered in inches of mercury (Hg). Log all types of precipitation, trace or more, and give the liquid equivalent of snow (one inch of snow equals one-tenth inch of rain). The relative humidity should be measured at the base station by psychrometer or other appropriate device. Enter the cloud cover at the time of observation using the descriptive terms defined in appendix C. Dew or frost should be entered as light, moderate, or heavy, depending on your observation of the general conditions around the base station in the morning. Soil temperature/moisture should be entered for the appropriate location and depth, as needed, during the growing season. Evaporation would be read once daily at the time of your first daily observation. The sunrise/sunset times, tides, and moon phases can be taken from calendars, newspapers, almanacs, etc.

In addition to entering the actual data above, you should make every effort to record the daily forecast. This will help you to compare the observations from your farm with the forecast and give you a good feel for which source is the best for your farm. Select two or three sources, such as the newspaper, local radio or television station, or your consulting meteorologist, and encode appropriately using the same guidelines for each variable as discussed above.

In the additional comments area at the bottom of the page, you should add any observations that might possibly prove to be of value to you or a professional meteorologist as you analyze the data at a later date. Add such data as unusual weather events, time of a marked wind shift, interesting natural or animal signs or activities, solar radiation measurements, weather data from nearby farms, towns, or cities, plus anything else of potential interest or significance. Just below this section, it will be very helpful if you'll write the page number of the corresponding isohyetal/isothermal map.

Finally, you should faithfully and diligently record all significant data with regard to each major crop you produce. Note seed type, planting date, planting technique, type of fertilizer, all herbicide and pesticide applications, the amount and date of each irrigation application, each significant stage in the maturity of your crop(s), major problems encountered, and yields (by individual fields, if possible—see figures 16 and 17 in chapter five). This data should be kept with your weather records for it will be used to correlate or associate the weather with the development and maintenance of your crop(s).

To double the value and safeguard your priceless weather records, always make a carbon copy of every Daily Farm Weather Log and

FIGURE 62. DAILY FARM WEATHER LOG

DATE ______________ (day / month / year)   BASE STATION ______________

CURRENT WEATHER OBSERVATIONS

TIME	MAXIMUM TEMP. °F	MINIMUM TEMP. °F	WIND	PRESSURE INS. HG	PRECIP. INCHES	RELATIVE HUMIDITY %	CLOUD COVER

DEW ______________ FROST ______________

SUNRISE ______________

SUNSET ______________

HIGH TIDES ______________ ______________

LOW TIDES ______________ ______________

MOON PHASE ______________

EVAPORATION (HUNDREDTHS OF AN INCH) ______________

LOCATION	SOIL TEMP. SOIL MOIST.	DEPTH	DEPTH

FORECAST

WEATHER	SOURCE	SOURCE	SOURCE
MAX. TEMP. °F			
MIN. TEMP. °F			
PRECIP. PROB.			
TYPE/AMT. PRECIP.			
WIND SPEED/DIR.			
CLOUD COVER			

ADDITIONAL COMMENTS

Note: Corresponding Isohyetal/Isothermal Map, page ______

Isohyetal/Isothermal Map. This will also save countless hours of needless transcribing of your data, should you mail your records to a consulting meteorologist for analysis.

If every farmer in America kept a set of records as described, it would literally be worth billions to agriculture. It would provide a truly dense network of quality weather data to enable professional meteorologists to make the kind of climate analyses that are really needed for a proper assessment of the effect of weather on American agriculture and, of course, your farm.

APPENDIX C COMMONLY USED TERMS FOR WEATHER OBSERVATIONS AND FORECASTS

You'll find it much easier to interpret the forecasts you hear, see, or read and to encode the daily observations from your Farm Weather Network if you're familiar with the terms most frequently used by meteorologists. Both official and popular terms are given below because they are used interchangeably by most forecasters and observers. This often confuses the layman who's not aware of the similarity or differences between the respective terms.

In reviewing the sky conditions, keep in mind that they normally refer to the presence or absence of a certain amount of thick or opaque clouds which completely obscure the sun. These clouds will generally occur at the lower and middle levels (see appendix D) but, occasionally, the higher clouds will also be opaque enough to qualify.

FAIR—This generally means pleasant weather conditions with no rain, very few low clouds, good visibility, and light winds. Official weather observers will often call the weather fair when there is a complete cloud cover of high cirrus clouds (see appendix D) yet the sun is still visible through the clouds. This is an official term used by the National Weather Service.

CLEAR—Officially, this means that clouds cover less than ten percent of the sky.

SUNNY—This is not an official term but is popularly used to mean the same as clear.

MOSTLY SUNNY—This is not an official term but is popularly used to mean less sunshine than clear or sunny but more sunshine or less clouds than partly cloudy.

PARTLY CLOUDY—Popularly, this means that there are quite a few clouds around but not enough to obscure the sun or sky completely at any moment. Officially, it refers to an observed or forecasted cover of low clouds over as much as 30 percent but no more than 60 percent of the sky.

MOSTLY CLOUDY—This is not an official term but it is popularly used to mean more cloud cover than partly cloudy but less cloud cover than cloudy.

CLOUDY—This is an official term which in popular usage refers to the state of weather when clouds almost totally prevail at the expense of the sun or completely obscure the stars at night. In official forecast/observation terminology it refers to a cloud cover of 70 percent or more.

VISIBILITY—The farthest distance in a given direction that you can identify a prominent object with the unaided eye during the day or clearly see a bright light at night. Officially, the prevailing visibility at an airport would be resolved into a single value consisting of the average visibility around the horizon circle.

PRECIPITATION—Any or all forms of liquid or solid water particles that fall to the ground. *Note:* Precipitation forecasts are usually expressed in terms of percentages. As explained in chapter one, a 50 percent chance of rain is supposed to mean that it almost certainly will rain but only over half of a given area. The actual amount of rain expected is rarely given. If you'll substitute the words "of the area" for "chance of," you'll have the correct meaning of the precipitation forecast. For example, a 30 percent chance of rain would read: showers will almost certainly occur over 30 percent of the area.

RAIN—Officially refers to liquid water drops with a diameter of 0.5 millimeters or greater.

VERY LIGHT RAIN—Does not completely wet an exposed surface regardless of duration.

LIGHT RAIN—The rate of fall varies between a trace and a tenth of an inch per hour with no more than one-hundredth in six minutes.

MODERATE RAIN—The rate of fall varies between eleven-hundredths and thirty-hundredths inch per hour with no more than three-hundredths in six minutes.

HEAVY RAIN—The rate of fall is greater than thirty-hundredths inch an hour or three-hundredths inch in six minutes.

RAIN SHOWER—Rain which falls from a convective cloud (one that has formed as a result of vertical motion in the atmosphere). A rain shower is usually a short-lived phenomenon with rapid changes in intensity as it moves overhead.

THUNDERSTORM—Almost always a cumulonimbus cloud (see appendix D) which results in thunder and usually visible lightning (lightning must be present to form thunder, but it's not always visible), locally heavy rain, gusty winds, and sometimes hail. Thunderstorms are a local phenomenon, since the cumulonimbus cloud will rarely be more than ten to twenty miles in diameter.

DRIZZLE—Consists of water droplets less than 0.5 millimeter in diameter. The droplets are not much larger than fog droplets but the difference is that they do fall to earth.

SNOW—Precipitation which consists of white or translucent ice crystals often combined into snowflakes. A general rule of thumb is that one inch of snow is equal to one-tenth of an inch of rain.

SLEET—Raindrops or largely melted snowflakes that fall through a freezing layer of air near the earth's surface.

HAIL—Precipitation in the shape of balls or nonuniform lumps of ice, always formed by convective clouds.

SNOW SHOWER—Snow which falls from a convective cloud. A snow shower is usually a brief phenomenon with rapid changes in intensity as it moves overhead.

FREEZING RAIN—Rain which freezes on contact with an object at the earth's surface and results in a glazed coating.

FREEZING DRIZZLE—Drizzle which freezes on contact with an object at the earth's surface and results in a glazed coating.

FOG—Visible minute water droplets suspended in the atmosphere near the earth's surface.

GROUND FOG—Fog that hides less than sixty percent of the sky. Fog that covers the earth's surface to a height of less than six feet is called "shallow fog."

HAZE—Minute dust or salt particles which reduce horizontal visibility. This is called dry haze, but when water collects on the minute haze particles it becomes "damp haze" or "mist."

APPENDIX D
CLOUD TYPES

In certain sky conditions listed in appendix C the word "cloudy" generally refers to low or middle clouds. However, these sky condition terms are very loosely applied and often tend to confuse the issue rather than clarify it. The confusion results from the fact that there are three main categories of clouds, each of which is capable of obscuring the sun. Yet the sky condition terms in forecasts give no hint as to which type of cloud is responsible for the loss of sunlight. This is significant, particularly to the meteorologist, for it makes a great deal of difference if the "partly cloudy" conditions are a result of high, middle, or low clouds. That's why the hourly reports of weather (generally seen only by pilots and meteorologists) from official observation sites across the country describe each individual layer of clouds overhead and what percentage of the sky it obscures.

The significance of knowing the altitude of clouds overhead is that each type of cloud is formed differently and the method of formation is significant to the type of weather to be expected. Basically there are two types of clouds according to formation: those formed by rising air currents, called *cumulus* (meaning accumulated or piled up), and those formed when water vapor in the air condenses into a visible cloud shape without benefit of upward motion. These are called *stratus* (layered or sheetlike).

According to international cloud classification, clouds are observed in three height groups: (1) Low, (2) Middle, and (3) High. The average height varies with latitude but in the middle latitudes the average heights will vary from the surface to 6,500 feet for Low, from 6,500 to 23,000 feet for Middle, and from 16,500 to 45,000 feet for High. It should also be noted that there is essentially one additional unofficial height classification referred to as the "towering clouds," where the bases can range from a few hundred feet above the earth's surface to well above 80,000 feet. The cumulonimbus cloud, or thunderstorm cloud, is officially a Low cloud yet could also be classified with the "towering" group.

The internationally accepted cloud classification also lists clouds according to genera, the main characteristics of clouds. Ten genera have been defined. They include cirrus, cirrocumulus, cirrostratus, altocumulus, altostratus, nimbostratus, stratocumulus, stratus, cumulus, and cumulonimbus. In the groupings below, certain genera are arbitrarily grouped in a specific height category even though that's not the only level at which they occur. Altostratus is often grouped with the High clouds, as are the tops of cumulonimbus. Cumulus and cumulonimbus are often grouped with the Middle clouds. Nimbostratus can be grouped with either Middle or Low clouds.

Being able to identify the following cloud types will make you better informed about the weather. Once you know the difference, you'll almost always have a reasonable idea about what type of weather to expect and how soon to expect it. Local variations of each type will occur and each may have some significance to the weather in your area.

High clouds usually move overhead coincident with a ridge of high pressure in the upper atmosphere. Middle clouds generally follow and will occasionally result in light rain, especially in winter. The last to move overhead are usually the Low clouds and their arrival frequently means rain. Hence, the convenient rule of thumb: High, Middle, Low, Rain, i.e., High clouds followed by Middle and Low clouds usually result in rain.

HIGH CLOUDS

High clouds are almost entirely composed of ice crystals. Their heights range from 16,500 to 45,000 feet but the average height of the base of the clouds is between 20,000 and 30,000 feet. There are three main types of High clouds: cirrus, cirrocumulus, cirrostratus.

Cirrus clouds are composed entirely of ice crystals and have a thin, wispy, feathery appearance. In mid-latitudes, cirrus clouds generally form at or above 25,000 feet, where below-freezing temperatures prevail. This type of cloud is often seen in the familiar "mares' tails" pattern, or well-defined wisps of cloud thicker at one end than at the other (see figure 63).

Cirrocumulus clouds appear as thin, white patches of clouds with a rippled appearance; they will appear as little cloudlets in a pattern. These clouds generally form at between 20,000 and 25,000 feet and consist of either highly supercooled water droplets or small ice crystals or both.

Figure 63. Cirrus above with Scattered Altostratus below

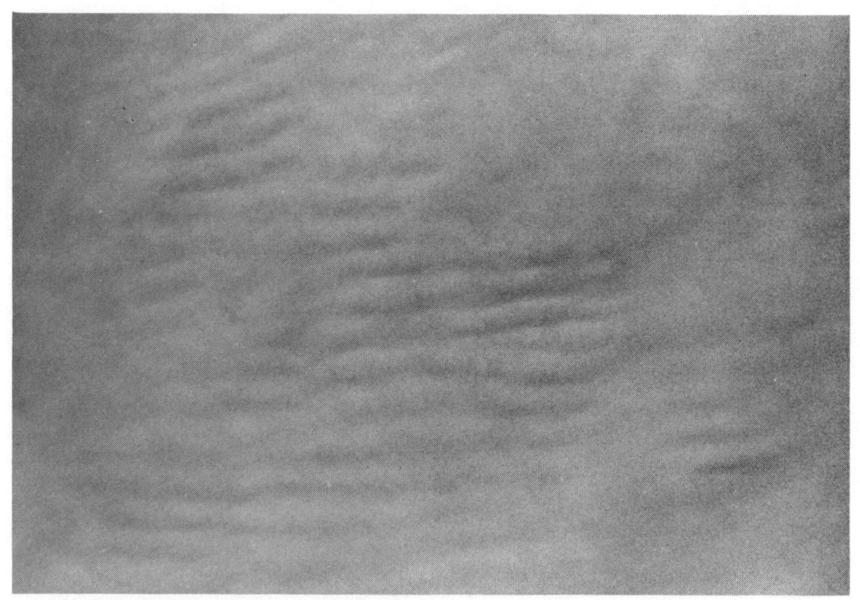

FIGURE 64. RIDGES OF CIRROCUMULUS OVERLAIN WITH MISTY CIRROSTRATUS

These are the clouds that form a "mackerel sky" (see figure 64).

Cirrostratus clouds are made entirely of ice crystals, yet form at the same altitudes as cirrocumulus clouds. They appear as thin sheets or a whitish veil and often cover the sky. These are the clouds which form halos around the sun and moon (see figures 65 and 66).

MIDDLE CLOUDS

Middle clouds can be composed of ice crystals, snowflakes, supercooled water droplets, or just ordinary water droplets. At middle latitudes, their bases generally average around 10,000 feet above the surface with average heights ranging from 6,500 to 23,000 feet. These clouds are either stratus- or cumulus-like: altostratus, altocumulus, and nimbostratus.

Altostratus clouds (see figure 67) are generally in dense sheets of gray or blue. The clouds are very uniform in appearance and often cover the entire sky, perhaps encompassing several thousand square miles at once. They are several hundred to several thousands of feet thick and, therefore, usually have a very complex composition consisting of ice crystals (near the top), ice crystals and/or snowflakes (in the middle), and supercooled or ordinary water droplets (near the bottom). Altostratus clouds are hard to distinguish from cirrostratus but the best rule of thumb is that altostratus does not create a halo effect, is thicker than cirrostratus, and appears to move more quickly.

Altocumulus clouds (see figure 68) are white or gray in appearance and occur as patches or layers of puffy clouds in a waved aspect. Altocumulus resembles cirrocumulus, but the latter never have shadows of their own and look smaller. These clouds do not create a halo effect. Altocumulus are invariably comprised of small liquid water droplets which give them their sharp outline.

The nimbostratus cloud is referred to as the "rain cloud," for precipitation falling from the cloud reaches the ground in most cases. It is darker than stratus and altostratus and is gray colored or dark. This cloud is composed of suspended water droplets, sometimes supercooled, and of falling raindrops and/or snowflakes or crystals. Nimbostratus is always accompanied by virga, rain which falls from the cloud in a funnel shape but never reaches the ground. (Virga is often confused by the layman with the funnel cloud of a tornado.) Fractostratus, or low scud clouds, often accompany nimbostratus if the wind is strong. The nimbostratus cloud is a tall cloud and usually rises several thousand feet in the atmosphere; it usually results from a thickening of altostratus, which is why it was grouped with the Middle clouds in this appendix. It could also have been grouped with the towering clouds of cumulonimbus and, sometimes, cumulus.

Note: In actual practice it is extremely difficult for the layman to tell the difference between Middle and High clouds. Even trained observers can confuse altostratus with cirrostratus or altocumulus with cirrocumulus. There are specific differences as discussed above, but it is often next to impossible to tell precisely the height of Middle and High clouds unless they have actually been penetrated by plane or weather balloon.

LOW CLOUDS

Low clouds range in height from just above the earth's surface to 6,500 feet. Their composition varies considerably because of the markedly different cloud types, especially when cumulonimbus, the thunderstorm cloud, is included. There are four main types of Low clouds, according to the international classification: stratus, stratocumulus, cumulus, and cumulonimbus.

Stratus clouds (see figure 69) have gray uniform bases; they resemble fog with the base above the ground. In fact, stratus often develops from fog that has risen above the earth's surface. Stratus is usually the lowest of all clouds and forms from as low as a hundred feet above the ground to several thousand feet. Stratus usually consists of relatively widely-dispersed water droplets. Since there is little vertical development, the only

Figure 65. Cirrostratus

Figure 66. Cirrostratus

Figure 67. Altostratus

FIGURE 68. ALTOCUMULUS

FIGURE 69. STRATUS ALONG BASE OF MOUNTAIN WITH ALTOSTRATUS ABOVE

FIGURE 70. STRATOCUMULUS

FIGURE 71. CUMULUS

type of precipitation associated with true stratus clouds is a very fine drizzle. Stratus is often confused with stratocumulus (see figure 70) and nimbostratus but it is grayer and more uniform at the base than the former and doesn't have the rain associated with the latter.

Stratocumulus clouds (see figure 70) are composed of small water droplets, occasionally accompanied by soft hail, larger droplets, and snowflakes. They frequently form in clear air and from the rising of stratus or transformation of stratus or nimbostratus. Stratocumulus clouds can be gray or whitish but, unlike stratus, have a rounded appearance and nonuniform bases. They do not produce rain but often transform into rain, or nimbostratus, clouds.

Cumulus clouds (see figure 71) appear as rising domes or often as puffy, white clouds with a cauliflower-like appearance. It is composed of the entire spectrum of liquid to frozen water and can result in any type of precipitation possible from all other cloud types. The shape of cumulus clouds is constantly changing due to their formation, which results from rising currents of air from below and horizontal winds above the earth's surface. They usually mean fair weather unless they begin to mass together and/or show a great deal of vertical development. They are usually greater in height than either stratus or stratocumulus.

Cumulonimbus clouds (see figures 72 and 73) are the most spectacular and awesome of all ten genera. These are the thunderstorm clouds which characteristically form into the classic anvil-like thunderhead several tens of thousands of feet above the earth's surface. The bases of cumulonimbus are usually within the Low cloud range, but the upper portions of the cloud extend into and above either the Middle or High clouds. Their height is usually determined by the height of the troposphere (lower layer of the atmosphere) which ranges in the summertime from 20,000 to 30,000 feet over southern Canada to over 80,000 feet along the Gulf of Mexico. The classic anvil-like shape results when the top of the cloud meets the lower portion of the stratosphere and flattens out in the warmer air. This usually marks the final stage of the thunderstorm before it begins to dissipate. As with cumulus, cumulonimbus is composed of the full spectrum of water in all its phases and the results are usually exciting, with locally heavy rain, hail, gusty winds, lightning, and often tornadoes. As the cumulonimbus cloud advances toward you there is usually a sudden increase in pressure as the cold, dense air from many thousands of feet above the surface rushes toward you. Tornadoes can theoretically occur in any cumulonimbus cloud but have been shown to be more frequent in the smaller, more sharply defined thunderstorm clouds which are generally associated with a cold front. Cumulonimbus clouds are positively the most dangerous of all

Figure 72. Cumulonimbus

FIGURE 73. CUMULONIMBUS

FIGURE 74. CUMULUS HUMULIS

FIGURE 75. CUMULUS FRACTUS

FIGURE 76. CUMULUS CONGESTUS

FIGURE 77. CUMULUS CONGESTUS

types and a definite hazard to all aircraft, regardless of size.

The above groupings constitute the major classification, or genera, of clouds according to height and composition. Less well-known, yet equally as interesting, cloud types are shown below in no particular order. They are arranged more for their novelty or beauty than for their significance to any particular type of weather.

Cumulus humulis clouds are fair-weather cumulus (see figure 74).

Cumulus fractus clouds are often associated with severe weather (see figure 75).

Cumulus congestus clouds are rain shower clouds, one step above cumulus and one step below cumulonimbus (see figures 76 and 77).

APPENDIX E PRACTICAL OBSERVATIONS ABOUT WEATHER

For thousands of years man has recorded his observations about weather. He has used his practical knowledge of weather to sail the seas, produce crops, protect his nation, catch fish, harness energy, protect himself from harm, and hunt for food.

A knowledge of wind and sea currents helped the ancient Phoenicians sail the Mediterranean Sea. Farmers have relied on their observations of weather to time their planting and predict their yields since man first learned how to grow his own food. The Russians wisely withdrew as the armies of Napoleon and Hitler advanced and let the harsh Russian winter do their fighting for them. A good fisherman knows the effect of weather on fish—e.g., before a storm, fish come to the surface to feed on insects flying closer to the water because of lower atmospheric pressure. Sunlight and wind have long been sources of power and heat—e.g., how much electrical energy would it take to drive all the windmills in the world? Indians knew better than to pitch their tents at the bottom of a mountain ravine along the east slope of the Rockies, yet modern buildings are frequently damaged by Chinook or mountain wave winds. Since the first hunter stalked game, hunters have observed that animals, like insects, react to changes in weather.

The following is by no means a complete listing of practical observations but you may find them useful as well as interesting. "Old-timers" in your area usually have a good idea of the significance of changing weather. See if they agree with the observations listed below.

1. A south wind means warmer temperatures (warm air moving northward ahead of a cold front).
2. Rain or snow is coming when:
   - low clouds move in behind middle and high clouds (especially if barometer is falling).
   - there is a ring around the moon. (High cirrus clouds, through which the moon is still visible, create the halo; middle and low clouds usually follow.)
   - clouds begin to develop upwards. (Vertical motion will condense moisture to form larger cloud; raindrops will form and when they're big enough they'll literally fall to earth.)
   - a dark threatening sky is observed to the west or northwest. (A cold front may be approaching.)
   - clouds move ever faster from the west or southwest and the wind gets ever stronger from the south. (A cold front is approaching.)
3. Clear skies are near when:
   - dark clouds become lighter and steadily rise in altitude. (The cold front has already moved on to the east.)
   - the wind shifts from south or southeast to west or northwest. (A cold front has just moved past your location.)
   - pressure rises rapidly.
4. A north wind means falling temperatures. (Cold air is moving in behind the cold front.)
5. Dew will not form on a cloudy night. The cloud cover keeps the earth from losing heat; therefore, the temperature near the earth's surface cannot cool off enough to condense the moisture in the air and form dew.
6. Fog forming on a pond in autumn or spring usually warns of frost. If the temperature has fallen low enough by evening to form fog, it will probably continue to fall until it freezes.
7. Rain is likely when leaves show their undersides. A strong south wind in advance of a cold front (which usually results in rain) will flip the leaves over.
8. Ropes tighten before a storm because of the higher humidity associated with warmer, moister air flowing northward in advance of a cold front.
9. The earth is closer to the sun in winter than summer (in the Northern Hemisphere). But it's colder in winter because the sun's rays are less direct.
10. Even if it's 38°F outside frost can form on the grass because the temperature can be freezing at grass level. Official temperature readings

are made at an elevation of five feet above the surface; the temperature will be cooler near the earth's surface since the atmosphere, on a clear night, cools from the ground up.

Some practical observations from nature:

1. Frogs croak more before a storm because the air is more humid and allows them to stay out of the water longer without drying their skin. Southerly flow in advance of a cold front usually brings more humid and warmer air northward.

2. Your joints are more likely to ache before a large storm. The low atmospheric pressure in advance of a weather system allows the gas in your joints to expand and create pain. As higher pressure moves in behind the weather system, it contracts the gas in your joints and the pain subsides. Arthritics or people who have had broken bones or major surgery are most susceptible to this affliction. Pain induced by a change in the weather is called a meteorpathic reaction.

3. Bats and birds fly lower before a storm. Their ears are sensitive to changes in atmospheric pressure, and low pressure in advance of a storm creates pain. So the lower they fly, the less the pain.

4. Bees return home before a storm because they sense the lower atmospheric pressure which generally precedes the rain.

5. Insects display more activity before a storm because the warmer air in advance of the weather front is more comfortable to them.

6. Spiders spin shorter, thicker webs in advance of stormy weather. They sense the low pressure in advance of rainy, windy weather. If their webs are long and thin, expect good weather for several days.

APPENDIX F
GLOSSARY

AIR MASS. A large body of air with a relatively uniform composition, which originates over one of several specific areas of the earth.

AMS. American Meteorological Society.

ANEMOMETER. A weather instrument designed to measure wind speed.

ANTICYCLONIC. A clockwise flow of air in the atmosphere; usually synonymous with high or high pressure system.

ATMOSPHERE. The envelope of air surrounding the earth which is held in place by the force of gravity. Most commonly observed weather phenomena occur in the troposphere (lower layer of the atmosphere).

ATMOSPHERIC PHENOMENA. Any type of weather occurrence actually observed in the atmosphere, such as clouds, wind, temperature, precipitation, etc.

ATMOSPHERIC PRESSURE. The pressure exerted upon a single point by the weight of the atmosphere above that point.

BAROMETER. A device for measuring atmospheric pressure.

BASE STATION. A collection of weather instruments designed to be a reference point for all other weather instruments distributed in a Farm Weather Network. There can be more than one base station per farm.

Basis. The difference between the futures market price and the local cash bid for the same delivery period.

BCD output. Binary-coded-decimal output from a weather instrument which can be fed directly into a computer for instant analysis or transmitted to another location in either binary, coded, or decimal format.

Bermuda High. A commonly observed atmospheric phenomenon in the North Atlantic Ocean which results in a large, semi-permanent high over the eastern and southeastern U.S.

Cash market. The grain elevator, cotton merchant, or livestock buyer in the local area where you trade.

Cash merchant. The end recipient of the raw product who either processes the product as well or sells it to others who will process the product accordingly.

CCM. Certified consulting meteorologist. An official designation of the American Meteorological Society given only after rigorous testing.

Climate variability. The variation of climate (long-term weather patterns) over periods of time ranging from months to years.

Cold air drainage. An atmospheric phenomenon where cold air sinks to the lowest places of the local topography, such as valleys, ponds, river and stream beds.

Cold front. Essentially the leading edge of a cold air mass or a non-occluded front which moves so as to replace warmer air with cold.

Commercial meteorology. Often called industrial meteorology, it refers to the application of meteorological information and techniques to commercial or industrial problems.

Consulting meteorologist. An experienced professional meteorologist who possesses several years of forecasting experience plus several years of practical application of his knowledge in a special area.

Crop-yield model. A statistical association, between critical weather variables (temperature, precipitation, soil moisture, and soil temperature, etc.) and crop maturity and resultant yields, constructed in order to estimate expected yields at any point in the growing season, given the actual weather conditions up to that point and assuming a given pattern of weather through the remainder of the growing season.

CYCLONIC. Counterclockwise flow of air in the atmosphere; usually synonymous with low or low pressure system.

DAILY CLIMATOLOGY. A long-term average of maximum and minimum temperatures and the frequency of precipitation for any given day of the year; usually constructed a month at a time to show the normal pattern of weather for that month.

"DRY" FIELD. A growing area which receives rain only infrequently and will be drier than other growing areas nearby. Compare to "wet" field.

EVAPORATION PAN. A pan used to measure the rate of evaporation of water into the atmosphere; part of a set of instruments designed to assist in irrigation scheduling.

EVAPOTRANSPIRATION. The process by which water is lost to the atmosphere by evaporation of liquid or solid water and by transpiration from plants.

FARM WEATHER NETWORK. A complete set of weather instruments, strategically situated about a farm, designed to monitor the effect of weather on a farm.

FIFTY PERCENT CHANCE OF SHOWERS. Correctly used, it refers to the almost certain possibility that showers will occur, but only over 50% of a given area, during the time span referred to in the forecast.

FRONT. In meteorology, the transition zone or interface between two air masses; usually characterized by a trough of low pressure, a change in wind direction, humidity, and temperature.

FRONTAL SYSTEM. A system of fronts (warm, cold, occluded) as depicted on a weather map.

FUTURES MARKET. A commodity exchange.

GERMINATE. To cause to sprout or grow.

GLACIAL. An interval of geologic time characterized by the equatorward advance of ice during an ice age.

GLOBAL CLIMATE. A composite global average of temperature and precipitation.

GREENWICH MEAN TIME. Twenty-four-hour universal time centered on 0° longitude at Greenwich, England.

HEAT ISLAND. A temperature phenomenon commonly observed in cities where a specific area within the city will consistently show higher temperature readings than the surrounding areas.

HEDGER. One who makes a forward purchase or sale, through a commodity exchange, the purpose of which is to lock in price.

HIGH. An area of high pressure denoted by a capital "H" on a weather map. It is always associated with anticyclonic circulation.

HIGH PRESSURE CENTER. The center of a high pressure system; point of maximum pressure.

HIGH PRESSURE SYSTEM. Synonymous with HIGH.

HUMIDITY. A measure of the amount of water vapor in the air.

HYGROMETER. An instrument or device for measuring the water vapor content of the atmosphere.

ICE AGE. A major interval of geologic time characterized by extensive sheets of ice which formed periodically over several areas of the world.

INFRARED RADIATION. Long-wave radiation emitted by the earth at night.

ISOHYETS. Refers to lines of equal rainfall, drawn on the isohyetal/isothermal map. Drawing these lines allows you to estimate rainfall between gauges.

ISOTHERMS. Refers to lines of equal temperature value, drawn on the isohyetal/isothermal map. Drawing these lines allows you to estimate temperature between thermometers.

JET STREAM. Strong winds confined to a relatively narrow stream in the atmosphere.

LITTLE ICE AGE. A period from the 1600s through the 1800s when the extent of the glaciers over the Northern Hemisphere approached the size of the last major glacial period.

LOCAL PROFESSIONAL. A professional commodity trader, a member of the exchange, who deals (for his clients only) in the commodities market.

LONG-RANGE FORECASTS. Forecasts for periods in excess of five days.

LOW. An area of low pressure denoted by a capital L on a weather map. Always associated with cyclonic flow.

LOW PRESSURE CENTER. Synonymous with LOW.

MAXIMUM/MINIMUM THERMOMETER. A thermometer designed to measure both the maximum and minimum temperature in a given interval of time.

METEOROLOGIST. Literally, one who studies atmospheric phenomena. Used in this book to refer to one who possesses at least a BS degree in meteorology.

MID-LATITUDE JET. A jet-stream wind which flows from west to east (Northern Hemisphere) at mid-latitudes. This is usually the dominant jet stream and weather maker over the U.S.

NATURAL PHENOMENA. Observed habits or activities of plants and animals.

NOAA. National Oceanic and Atmospheric Administration. An organization within the U.S. Department of Commerce, NOAA is the parent organization for the National Weather Service.

NORMAL. Used in meteorology to mean the same as average.

OCCLUDED FRONT. A composite front formed when a cold front overtakes a warm front or stationary front.

PLUVIAL. An interval of geologic time characterized by abundant rainfall, usually in the lower latitudes, and associated with the glacial advances.

PRESSURE. In meteorology, usually synonymous with ATMOSPHERIC PRESSURE.

PROFESSIONAL FORECASTER. An experienced weather forecaster with at least three to four years of practical experience; in actual practice this status is usually applied to those individuals with 10 or more years of experience; (this is the basic requirement for a non-degreed weather forecaster to be accepted as a full member of the AMS).

PSYCHROMETER. A device or instrument for measuring the amount of water vapor in the atmosphere.

RADAR SUMMARY MAP. A weather map which depicts a summary of radar observations across the U.S.

RADIOSONDE. An instrument, carried aloft by a balloon, designed to measure humidity, temperature, and pressure. By tracking this instrument you can also determine wind speed and direction.

RAIN GAUGE. An instrument for measuring the amount of rainfall.

RAIN TRACK. Definable patterns of consistently higher than normal rainfall in a given area.

RELATIVE HUMIDITY. Popularly called HUMIDITY and expressed as a per cent; for practical purposes it is synonymous with HUMIDITY.

SHORT-RANGE FORECAST. Used in this book to refer to a weather forecast for a period of no more than five days and usually less than three.

SOIL BLOCKS. Electronic sensors, buried in the soil to an appropriate depth, designed to measure the amount of moisture in the soil at that depth.

SOIL MOISTURE. In this book, used to refer to moisture within the root zone of plants.

Solar cycle. Time between a maximum and minimum occurrence of sunspots.

Solar radiation. The complete spectrum of electromagnetic radiation emitted by the sun.

Speculator. Anyone who participates in a commodity market who is not a consumer, producer, or maker of the actual cash transaction.

Stand. Used in this book to refer to the young seedling just after it has broken ground and is no more than two to three inches tall.

Stationary front. Often referred to as a "quasi-stationary" front because it is moving at a very slight speed—less than about five miles per hour.

Statistical outlook. A mathematical projection of the temperature or precipitation for months or years in advance.

Statistical projection. Synonymous with statistical outlook.

Stratosphere. The next layer of the atmosphere above the troposphere, which is the lowest layer.

Sunspot. A dark spot observed on the surface of the sun.

Surface map. Synonymous with weather map, which depicts highs, lows, fronts, clouds, precipitation, temperature, winds, and pressure.

Temperature. A measure of the degree of heat or cold on a given scale.

Thermograph. A self-recording thermometer which can be used to measure both air and soil temperature.

Thermometer. A device or instrument for measuring temperature.

Troposphere. The lowest layer of the atmosphere in which most of the observed weather phenomena occur.

Upper air. That portion of the atmosphere above the lower troposphere.

UPPER-ATMOSPHERIC DATA. Measured weather variables, such as wind, temperature, and humidity, above the surface of the earth. These are usually observed by radiosondes.

UPPER-LEVEL RIDGE. A ridge of high pressure in the upper air.

UPPER-LEVEL TROUGH. A trough of low pressure in the upper air.

UPPER-LEVEL WINDS. Often called "winds aloft." Refers to wind speeds and directions at a given level in the atmosphere above the earth's surface.

USDA. U.S. Department of Agriculture.

WARM FRONT. A non-occluded front in which warm air moves to replace cold air.

WEATHER MANAGEMENT. As applied to agriculture, refers to the complete process of monitoring and recording the effect of weather on a farm and using this information to save operating costs and to market crops.

WEATHER SYSTEM. An organized pattern of weather, synonymous with FRONTAL SYSTEM.

"WET" FIELD. A growing area which receives abundant rainfall and is usually wetter than surrounding areas. Compare to "dry" field.

WIND. Air that is in motion relative to the surface of the earth.

WIND DIRECTION. The direction from which the wind is blowing.

WIND VANE. A device or instrument used to determine wind direction or direction from which wind is blowing.

INDEX